T0206231

SpringerBriefs in Statistics

JSS Research Series in Statistics

The current research of statistics in Japan has expanded in several directions in line with recent trends in academic activities in the area of statistics and statistical sciences over the globe. The core of these research activities in statistics in Japan has been the Japan Statistical Society (JSS). This society, the oldest and largest academic organization for statistics in Japan, was founded in 1931 by a handful of pioneer statisticians and economists and now has a history of about 80 years. Many distinguished scholars have been members, including the influential statistician Hirotugu Akaike, who was a past president of JSS, and the notable mathematician Kiyosi Itô, who was an earlier member of the Institute of Statistical Mathematics (ISM), which has been a closely related organization since the establishment of ISM. The society has two academic journals: the Journal of the Japan Statistical Society (English Series) and the Journal of the Japan Statistical Society (Japanese Series). The membership of JSS consists of researchers, teachers, and professional statisticians in many different fields including mathematics, statistics, engineering, medical sciences, government statistics, economics, business, psychology, education, and many other natural, biological, and social sciences. The JSS Series of Statistics aims to publish recent results of current research activities in the areas of statistics and statistical sciences in Japan that otherwise would not be available in English; they are complementary to the two JSS academic journals, both English and Japanese. Because the scope of a research paper in academic journals inevitably has become narrowly focused and condensed in recent years, this series is intended to fill the gap between academic research activities and the form of a single academic paper. The series will be of great interest to a wide audience of researchers, teachers, professional statisticians, and graduate students in many countries who are interested in statistics and statistical sciences, in statistical theory, and in various areas of statistical applications.

More information about this subseries at http://www.springer.com/series/13497

Taka-aki Shiraishi · Hiroshi Sugiura ·
Shin-ichi Matsuda

Pairwise Multiple Comparisons

Theory and Computation

Taka-aki Shiraishi
Faculty of Science and Engineering
Nanzan University
Nagoya, Aichi, Japan

Hiroshi Sugiura
Faculty of Science and Engineering
Nanzan University
Nagoya, Aichi, Japan

Shin-ichi Matsuda
Faculty of Science and Engineering
Nanzan University
Nagoya, Aichi, Japan

ISSN 2191-544X ISSN 2191-5458 (electronic)
SpringerBriefs in Statistics
ISSN 2364-0057 ISSN 2364-0065 (electronic)
JSS Research Series in Statistics
ISBN 978-981-15-0065-7 ISBN 978-981-15-0066-4 (eBook)
https://doi.org/10.1007/978-981-15-0066-4

This Springer imprint is published by the registered company Springer Nature Singapore Pte Ltd.
The registered company address is: 152 Beach Road, #21-01/04 Gateway East, Singapore 189721, Singapore

Preface

Analysis of variance methods are commonly used statistical procedures in multi-sample models. However, in the analysis of variance, since the homogeneity of means is tested and the confidence regions of the mean vector are given by an interior of ellipses, specific comparisons of means are not drawn. Multiple tests and simultaneous confidence intervals specify differences in means. Therefore, fields including medicine, pharmacy, biology, and psychology use multiple comparison procedures for data analyses. Tukey (1953), Miller (1981), Hochberg and Tamhane (1987), Hsu (1996), and Bretz et al. (2010) are some technical books on multiple comparisons. The present book discusses progressive multiple comparisons.

The detailed discussion of this monograph focuses on all-pairwise multiple comparisons of means in multi-sample models. Closed testing procedures based on maximum absolute values of some two-sample t-test statistics and based on F-test statistics are introduced in homoscedastic multi-sample models. The results suggest that (i) multistep procedures are more effective than single-step procedures and the Ryan–Einot–Gabriel–Welsch (REGW) tests and (ii) confidence regions induced by multistep procedures are equivalent to simultaneous confidence intervals. Next, we introduce the multistep test procedure superior to the single-step Games–Howell procedure in heteroscedastic multi-sample models. Under simple ordered restrictions of means, we also discuss closed testing procedures based on maximum values of two-sample one-sided t-test statistics and based on Bartholomew's statistics. Furthermore, we introduce distribution-free procedures. Simulation studies are performed under the null hypothesis and some alternative hypotheses. Although single-step multiple comparison procedures are utilized in general, the closed testing procedures stated in the present book are fairly more powerful than the single-step procedures. In order to execute the multiple comparison procedures, the upper 100α percentiles of the complicated distributions are required. Classical integral formulas such as the Simpson's rule and the Gaussian rule have been used for the calculation of the integral transform that appears in statistical calculations. However, these formulas are not effective for the complicated distribution. As a numerical calculation, the authors introduce the Sinc method which is the optimum in terms of accuracy and computational cost.

Shiraishi writes Chaps. 1–4 about multiple comparison procedures. These chapters are translated into English from Japanese of Shiraishi and Sugiura (2018, Kyoritsu Shuppan Co., Ltd.). We obtain the permission of Kyoritsu Shuppan Co., Ltd. and publish these chapters. Matsuda writes computer simulations for comparing the simulated power of multiple comparison tests and statistical analysis of raw data in Chaps. 5 and 6, respectively. Sugiura writes Chap. 7 about computation of distribution functions for statistics under simple order restrictions. This chapter is translated into English from Japanese of Shiraishi and Sugiura (2015, J. Japan Statistical Society; Japanese Issue). We obtain the permission of Japan Statistical Society and publish Chap. 7. Shiraishi writes Chap. 8 as related topics.

Nagoya, Japan
June 2019

Taka-aki Shiraishi
Hiroshi Sugiura
Shin-ichi Matsuda

Acknowledgements The authors are grateful to two referees for valuable comments. We would like to thank Prof. Takemitsu Hasegawa of the University of Fukui for careful reading and criticisms of the manuscript of Chap. 7. He also made helpful suggestions on improving the presentation of Chap. 7. This research was supported in part by a Grant-in-Aid for Co-operative Research (C) 18K11204 and 19K11870 from the Japanese Ministry of Education. We would like to thank Editage (www.editage.com) for English language editing.

References

Bretz F, Hothorn T, Westfall P (2010) Multiple comparisons using R. Chapman & Hall, London
Hochberg Y, Tamhane AC (1987) Multiple comparison procedures. Wiley, New York
Hsu JC (1996) Multiple comparisons-theory and methods. Chapman & Hall, London
Miller RG (1981) Simultaneous statistical inference, 2nd edn. Springer, Berlin
Shiraishi T, Sugiura H (2015) The upper $100\alpha^{\star}$th percentiles of the distributions used in multiple comparison procedures under a simple order restriction. J Japan Stat Soc. Japanese Issue **44**, 271–314 (in Japanese)
Shiraishi T, Sugiura H (2018) Theory of multiple comparison procedures and its computation. Kyoritsu-Shuppan Co., Ltd. (in Japanese)
Tukey JW (1953) The problem of multiple comparisons. The collected works of John W. Tukey (1994), volume VIII: multiple comparisons. Chapman & Hall, London

Contents

Chapter 1
All-Pairwise Comparisons in Homoscedastic Multi-sample Models

Abstract We consider multiple comparison procedures among mean effects in homoscedastic k-sample models. We propose closed testing procedures based on the maximum values of some two-sample t-test statistics and based on F-test statistics. The results reveal that the proposed procedures are more powerful than single-step procedures and the REGW (Ryan-Einot-Gabriel-Welsch) type tests.

1.1 Introduction

We consider homoscedastic k-sample models under normality. $(X_{i1}, \ldots, X_{in_i})$ is a random sample of size n_i from the ith normal population with mean μ_i and variance σ^2 $(i = 1, \ldots, k)$, that is, $P(X_{ij} \leq x) = \Phi((x - \mu_i)/\sigma)$, where $\Phi(x)$ is a standard normal distribution function. Further, X_{ij}'s are assumed to be independent. Then unbiased estimators for μ_i, overall mean $\nu = \sum_{i=1}^{k} n_i \mu_i / n$, and σ^2, respectively, are given by $\hat{\mu}_i = \bar{X}_{i\cdot}$, $\hat{\nu} = \bar{X}_{\cdot\cdot}$, and

$$V_E = \frac{1}{m} \sum_{i=1}^{k} \sum_{j=1}^{n_i} (X_{ij} - \bar{X}_{i\cdot})^2, \tag{1.1}$$

where $\bar{X}_{i\cdot} := (1/n_i) \sum_{j=1}^{n_i} X_{ij}$, $\bar{X}_{\cdot\cdot} := (1/n) \sum_{i=1}^{k} \sum_{j=1}^{n_i} X_{ij}$,

$$m := n - k, \text{ and } n := \sum_{i=1}^{k} n_i. \tag{1.2}$$

The ratio $F_t := \sum_{i=1}^{k} n_i (\bar{X}_{i\cdot} - \bar{X}_{\cdot\cdot})^2 / \{(k-1) V_E\}$ is used to test the null hypothesis of no treatment effects,

$$H_0 : \mu_1 = \cdots = \mu_k, \tag{1.3}$$

as follows. We reject H_0 at level α if $F_t > F_m^{k-1}(\alpha)$, where $F_m^{k-1}(\alpha)$ denotes the upper $100\alpha\%$ point of F-distribution with degrees of freedom $(k-1, m)$. For specified i, i' such that $1 \leq i < i' \leq k$, if we are interested in testing

T.-a. Shiraishi et al., *Pairwise Multiple Comparisons*,
JSS Research Series in Statistics,
https://doi.org/10.1007/978-981-15-0066-4_1

$$\text{the null hypothesis } H_{(i,i')} : \mu_i = \mu_{i'} \text{ vs. the alternative } H^A_{(i,i')} : \mu_i \neq \mu_{i'}, \tag{1.4}$$

we can use the two-sided two-sample t-test. In this chapter, we consider test procedures for all-pairwise comparisons of $\{$the null hypothesis $H_{(i,i')}$ vs. the alternative $H^A_{(i,i')} \mid (i,i') \in \mathscr{U}\}$, where

$$\mathscr{U} = \{(i,i') \mid 1 \le i < i' \le k\}. \tag{1.5}$$

Tukey (1953) and Kramer (1956) proposed single-step procedures as multiple comparison tests of level α. Shiraishi (2011) proposed closed testing procedures. Our findings suggest that (i) the proposed multistep procedures are more powerful than single-step procedures and the REGW (Ryan-Einot-Gabriel-Welsch) tests, and (ii) confidence regions induced by the multistep procedures are equivalent to simultaneous confidence intervals. The REGW test procedures are included in the SPSS system.

1.2 The Tukey–Kramer Method

We introduce two distribution functions of $TA(t)$ and $TA^*(t)$.

$$TA(t) := k\int_0^\infty \left[\int_{-\infty}^\infty \{\Phi(x) - \Phi(x - \sqrt{2}\cdot ts)\}^{k-1} d\Phi(x)\right] g(s|m)ds, \tag{1.6}$$

$$\begin{aligned} TA^*(t) :=& \sum_{j=1}^k \int_0^\infty \left[\int_{-\infty}^\infty \prod_{\substack{i=1\\ i\neq j}}^k \left\{\Phi\left(\sqrt{\frac{\lambda_{ni}}{\lambda_{nj}}}\cdot x\right)\right.\right. \\ &\left.\left. - \Phi\left(\sqrt{\frac{\lambda_{ni}}{\lambda_{nj}}}\cdot x - \sqrt{\frac{\lambda_{ni}+\lambda_{nj}}{\lambda_{nj}}}\cdot ts\right)\right\} d\Phi(x)\right] g(s|m)ds, \end{aligned}$$

where $\lambda_{ni} := n_i/n \ (i = 1,\ldots,k)$,

$$g(s|m) := \frac{m^{m/2}}{\Gamma(m/2)2^{(m/2-1)}} s^{m-1}\exp(-ms^2/2), \tag{1.7}$$

and m is defined in (1.2).

$TA(t/\sqrt{2})$ is referred to as studentized range distribution. We put

$$T_{i'i} := \frac{\bar{X}_{i'\cdot} - \bar{X}_{i\cdot}}{\sqrt{V_E\left(\frac{1}{n_i} + \frac{1}{n_{i'}}\right)}} \qquad ((i,i') \in \mathscr{U}). \tag{1.8}$$

Then we get Theorem 1.1.

Theorem 1.1 *For* $t > 0$,

$$TA(t) \leq P_0\left(\max_{(i,i')\in\mathscr{U}} |T_{i'i}| \leq t\right) \leq TA^*(t) \tag{1.9}$$

holds, where $P_0(\cdot)$ *stands for probability measure under the null hypothesis* H_0. *When* $n_1 = \cdots = n_k$ *is satisfied, both of the inequalities of (1.9) become an equality.*

The left-hand side of the inequality (1.9) is derived from main theorem of Hayter (1984). The right-hand side of the inequality (1.9) is given by Shiraishi (2006). For a given α such that $0 < \alpha < 1$, we put

$$ta(k, m; \alpha) := \text{a solution of } t \text{ satisfying the equation } TA(t) = 1 - \alpha. \tag{1.10}$$

[1.1] Single-Step Tests Based on t-Statistics

The Tukey–Kramer simultaneous test of level α for the null hypotheses $\{H_{(i,i')} \mid (i, i') \in \mathscr{U}\}$ consists in rejecting $H_{(i,i')}$ for $(i, i') \in \mathscr{U}$ such that $|T_{i'i}| > ta(k, m; \alpha)$. From the left inequality of (1.9), we find that the Tukey–Kramer simultaneous test is conservative. Under the condition of $\max_{1\leq i\leq k} n_i / \min_{1\leq i\leq k} n_i \leq 2$, Shiraishi (2006) found that the values of $TA^*(t) - TA(t)$ are nearly equal to 0 for various values of t from numerical integration. Therefore, the conservativeness of the Tukey–Kramer method is small.

Let us put, for $\boldsymbol{\mu} = (\mu_1, \ldots, \mu_k)$,

$$T_{i'i}(\boldsymbol{\mu}) := \frac{\bar{X}_{i'\cdot} - \bar{X}_{i\cdot} - (\mu_{i'} - \mu_i)}{\sqrt{V_E\left(\frac{1}{n_i} + \frac{1}{n_{i'}}\right)}} \quad ((i, i') \in \mathscr{U}). \tag{1.11}$$

Then we also get the following relation:

$$P\left(\max_{(i,i')\in\mathscr{U}} |T_{i'i}(\boldsymbol{\mu})| \leq t\right) = P_0\left(\max_{(i,i')\in\mathscr{U}} |T_{i'i}| \leq t\right). \tag{1.12}$$

[1.2] Simultaneous Confidence Intervals

From (1.9) and (1.12), we find that $100(1-\alpha)\%$ Tukey–Kramer simultaneous confidence intervals for all-pairwise $\{\mu_{i'} - \mu_i \mid (i, i') \in \mathscr{U}\}$ are given by

$$\mu_{i'} - \mu_i \in \bar{X}_{i'\cdot} - \bar{X}_{i\cdot} \pm ta(k, m; \alpha) \cdot \sqrt{V_E\left(\frac{1}{n_i} + \frac{1}{n_{i'}}\right)} \quad ((i, i') \in \mathscr{U}). \tag{1.13}$$

1.3 Closed Testing Procedures

We consider test procedures for all-pairwise comparisons of

$$\left\{\text{the null hypothesis } H_{(i,i')} \text{ vs. the alternative } H^A_{(i,i')} \,\middle|\, (i,i') \in \mathscr{U}\right\}.$$

Let us put

$$\mathscr{H} := \{H_{(i,i')} \mid (i,i') \in \mathscr{U}\}. \tag{1.14}$$

Then, the closure of $\mathscr{H}$ is given by

$$\overline{\mathscr{H}} = \left\{\bigwedge_{v \in V} H_v \,\middle|\, \emptyset \subsetneq V \subset \mathscr{U}\right\},$$

where $\bigwedge$ denotes the conjunction symbol (Refer to Enderton 2001). Then, we get

$$\bigwedge_{v \in V} H_v : \text{for any } (i,i') \in V,\ \mu_i = \mu_{i'} \text{ holds.} \tag{1.15}$$

For an integer J and disjoint sets $I_1, \ldots, I_J \subset \{1, \ldots, k\}$, we define the null hypothesis $H(I_1, \ldots, I_J)$ by

$$\begin{aligned} H(I_1, \ldots, I_J) : &\text{for any integer } j \text{ such that } 1 \le j \le J \\ &\text{and for any } i, i' \in I_j,\ \mu_i = \mu_{i'} \text{ holds.} \end{aligned} \tag{1.16}$$

From (1.15) and (1.16), for any nonempty $V \subset \mathscr{U}$, there exist an integer J and disjoint sets $I_1, \ldots, I_J$ such that

$$\bigwedge_{v \in V} H_v = H(I_1, \ldots, I_J) \tag{1.17}$$

and $\#(I_j) \ge 2$ $(j = 1, \ldots, J)$, where $\#(A)$ stands for the cardinal number of set A. For $H(I_1, \ldots, I_J)$ of (1.17), we set

$$M := M(I_1, \ldots, I_J) = \sum_{j=1}^{J} \ell_j,\ \ell_j := \#(I_j). \tag{1.18}$$

Let us put

$$T(I_j) := \max_{i<i',\ i,i' \in I_j} |T_{i'i}| \qquad (j = 1, \ldots, J).$$

Then, we propose the stepwise procedure [1.3].

[1.3] Stepwise Procedure Based on t-Statistics
For $\ell = \ell_1, \ldots, \ell_J$, we define $\alpha(M, \ell)$ by

$$\alpha(M, \ell) := 1 - (1 - \alpha)^{\ell/M}. \tag{1.19}$$

Corresponding to (1.6), we put

$$TA(t|\ell, m) := \ell \int_0^\infty \left[\int_{-\infty}^\infty \{\Phi(x) - \Phi(x - \sqrt{2} \cdot ts)\}^{\ell-1} d\Phi(x) \right] g(s|m) ds. \tag{1.20}$$

By obeying the notation $ta(k, m; \alpha)$, we get

$$TA(ta(\ell, m; \alpha(M, \ell))|\ell, m) = 1 - \alpha(M, \ell) = (1 - \alpha)^{\ell/M}, \tag{1.21}$$

that is, $ta(\ell, m; \alpha(M, \ell))$ is an upper $100\alpha(M, \ell)\%$ point of the distribution $TA(t|\ell, m)$.

(a) $J \geq 2$
Whenever $ta\left(\ell_j, m; \alpha(M, \ell_j)\right) < T(I_j)$ holds for an integer j such that $1 \leq j \leq J$, we reject the hypothesis $\bigwedge_{v \in V} H_v$.
(b) $J = 1$ $(M = \ell_1)$
Whenever $ta\,(M, m; \alpha) < T(I_1)$ holds, we reject the hypothesis $\bigwedge_{v \in V} H_v$.

By using the methods of (a) and (b), when $\bigwedge_{v \in V} H_v$ is rejected for any V such that $(i, i') \in V \subset \mathscr{U}$, the null hypothesis $H_{(i,i')}$ is rejected as a multiple comparison test.

Theorem 1.2 *The test procedure [1.3] is a multiple comparison test of level α.*

Proof It is trivial to verify that the level of the test procedure of (b) is α. We show that the level of the test procedure of (a) is α. Furthermore, we suppose without any loss of generality that H_0 is true and $\sigma^2 = 1$. $g(s|m)$ is a density function of $\sqrt{V_E}$. Since $\sqrt{V_E} \cdot T(I_1), \ldots, \sqrt{V_E} \cdot T(I_J)$ and $\sqrt{V_E}$ are independent, we get

$$\begin{aligned}
&P_0\left(T(I_j) \leq ta\left(\ell_j, m; \alpha(M, \ell_j)\right),\ j = 1, \ldots, J\right) \\
&= \int_0^\infty P_0\Big(T(I_j) \leq ta\left(\ell_j, m; \alpha(M, \ell_j)\right), \\
&\qquad\qquad j = 1, \ldots, J \;\Big|\; \sqrt{V_E} = s\Big) \cdot g(s|m) ds \\
&= \int_0^\infty P_0\Big(\sqrt{V_E} \cdot T(I_j) \leq s \cdot ta\left(\ell_j, m; \alpha(M, \ell_j)\right), \\
&\qquad\qquad j = 1, \ldots, J \;\Big|\; \sqrt{V_E} = s\Big) \cdot g(s|m) ds \\
&= \int_0^\infty \left\{ \prod_{j=1}^J P_0\left(\sqrt{V_E} \cdot T(I_j) \leq s \cdot ta\left(\ell_j, m; \alpha(M, \ell_j)\right)\right) \right\} \cdot g(s|m) ds.
\end{aligned} \tag{1.22}$$

Here by applying Corollary A.1.1 of Hsu (1996) to (1.22), it follows that

$$(1.22) \geq \prod_{j=1}^{J} \int_0^{\infty} P_0\left(\sqrt{V_E} \cdot T(I_j) \leq s \cdot ta\left(\ell_j, m; \alpha(M, \ell_j)\right)\right) \cdot g(s|m)ds. \tag{1.23}$$

Let us put

$$A(u|\ell) = \ell \int_{-\infty}^{\infty} \{\Phi(x) - \Phi(x - \sqrt{2} \cdot u)\}^{\ell-1} d\Phi(x).$$

Then from Hayter (1984), we find

$$P_0\left(\sqrt{V_E} \cdot T(I_j) \leq u\right) \geq A\left(u|\ell_j\right).$$

Here by using the relation

$$P_0\left(\sqrt{V_E} \cdot T(I_j) \leq s \cdot ta\left(\ell_j, m; \alpha(M, \ell_j)\right)\right) \geq A\left(s \cdot ta(\ell_j, m; \alpha(M, \ell_j))|\ell_j\right),$$

we have

$$\begin{aligned}
&\int_0^{\infty} P_0\left(\sqrt{V_E} \cdot T(I_j) \leq s \cdot ta\left(\ell_j, m; \alpha(M, \ell_j)\right)\right) \cdot g(s|m)ds \\
&\qquad \geq \int_0^{\infty} A\left(s \cdot ta(\ell_j, m; \alpha(M, \ell_j))|\ell_j\right) \cdot g(s|m)ds \\
&\qquad = TA\left(ta(\ell_j, m; \alpha(M, \ell_j))|\ell_j, m\right) \\
&\qquad = 1 - \alpha(M, \ell_j) \\
&\qquad = (1 - \alpha)^{\ell_j/M}.
\end{aligned} \tag{1.24}$$

By using (1.23) and (1.24), we get

$$\begin{aligned}
&P_0\left(\text{There exists } j \text{ such that } T(I_j) > ta\left(\ell_j, m; \alpha(M, \ell_j)\right)\right) \\
&\qquad = 1 - P_0\left(T(I_j) \leq ta\left(\ell_j, m; \alpha(M, \ell_j)\right),\ j = 1, \ldots, J\right) \\
&\qquad \leq 1 - \prod_{j=1}^{J} \left\{(1 - \alpha)^{\ell_j/M}\right\} \\
&\qquad = \alpha.
\end{aligned} \tag{1.25}$$

Therefore, the level of the test procedure of (a) for the null hypothesis $\bigwedge_{v \in V} H_v$ is α. □

From (1.17), we find

$$\overline{\mathscr{H}} = \Big\{ H(I_1, \ldots, I_J) \;\Big|\; \text{There exists } J \text{ such that } \bigcup_{j=1}^{J} I_j \subset \{1, \ldots, k\},\ \#(I_j) \geq 2\ (1 \leq j \leq J), \text{ and } I_j \cap I_{j'} = \emptyset\ (1 \leq j < j' \leq J,\ J \geq 2) \text{ are satisfied.} \Big\}$$

For $(i, i') \in \mathscr{U}$, let us put

$$\overline{\mathscr{H}}_{(i,i')} := \Big\{ H(I_1, \ldots, I_J) \in \overline{\mathscr{H}} \;\Big|\; \text{There exists } j \text{ such that } \{i, i'\} \subset I_j \text{ and } 1 \leq j \leq J \Big\}.$$

Then we get

$$\overline{\mathscr{H}} = \bigcup_{(i,i') \in \mathscr{U}} \overline{\mathscr{H}}_{(i,i')} \quad \text{and} \quad H_{(i,i')},\ H_0 \in \overline{\mathscr{H}}_{(i,i')}.$$

Therefore, by following (i) and (ii), we make a decision to reject or retain $H_{(i,i')}$ as a multiple comparison test of level α for $(i, i') \in \mathscr{U}$.

(i) Whenever all the elements of $\overline{\mathscr{H}}_{(i,i')}$ are rejected, $H_{(i,i')}$ is rejected.
(ii) Whenever there exists an element of $\overline{\mathscr{H}}_{(i,i')}$ that is not rejected, $H_{(i,i')}$ is not rejected.

For $k = 4$, all the elements $H(I_1, \ldots, I_J)$ of $\overline{\mathscr{H}}_{(1,2)}$ are stated in Table 1.1. From Table 1.1, in order to reject $H_{(1,2)}$ as a multiple comparison test, five null hypotheses must be rejected. Whenever the following (1-1)–(1-5) are satisfied, the closed testing procedure of [1.3] rejects $H_{(1,2)}$ as a multiple comparison test of level α (Table 1.2).

(1-1) $T(\{1, 2, 3, 4\}) = \max_{1 \leq i < i' \leq 4} |T_{i'i}| > ta(4, m; \alpha)$.
(1-2) $T(\{1, 2\}) = |T_{21}| > ta(2, m; \alpha(4, 2))$ or $T(\{3, 4\}) = |T_{43}| > ta(2, m; \alpha(4, 2))$.
(1-3) $T(\{1, 2, 3\}) = \max_{1 \leq i < i' \leq 3} |T_{i'i}| > ta(3, m; \alpha)$.
(1-4) $T(\{1, 2, 4\}) = \max\{|T_{21}|, |T_{41}|, |T_{42}|\} > ta(3, m; \alpha)$.
(1-5) $T(\{1, 2\}) = |T_{21}| > ta(2, m; \alpha)$.

From a definition, we can verify that $ta(\ell, m; \alpha) < ta(k, m; \alpha)$ holds for ℓ such that $2 \leq \ell < k$. For $\alpha = 0.05,\ 0.01$, we give the values of $ta(\ell; \alpha(M, \ell))$ in

Table 1.1 When $k = 4$, in testing the null hypothesis $H_{(1,2)}$ as a multiple comparison, the null hypotheses $H(I_1, \ldots, I_J) \in \overline{\mathscr{H}}_{(1,2)}$ that are tested as a closed testing procedure

M	$H(I_1, \ldots, I_J)$	J	ℓ
4	$H(\{1, 2, 3, 4\}) = H_0$, $H(\{1, 2\}, \{3, 4\}) : \mu_1 = \mu_2,\ \mu_3 = \mu_4$	$J = 1$, $J = 2$,	$\ell_1 = 4$ $\ell_1 = \ell_2 = 2$
3	$H(\{1, 2, 3\}) : \mu_1 = \mu_2 = \mu_3$, $H(\{1, 2, 4\}) : \mu_1 = \mu_2 = \mu_4$	$J = 1$, $J = 1$,	$\ell_1 = 3$ $\ell_1 = 3$
2	$H(\{1, 2\}) : \mu_1 = \mu_2$	$J = 1$,	$\ell_1 = 2$

Table 1.2 When $k = 5$, in testing the null hypothesis $H_{(1,2)}$ as a multiple comparison, the null hypotheses $H(I_1, \ldots, I_J) \in \overline{\mathscr{H}}_{(1,2)}$ that are tested as a closed testing procedure

M	$H(I_1, \ldots, I_J)$
5	$H(\{1, 2, 3, 4, 5\})$, $H(\{1, 2, 3\}, \{4, 5\})$, $H(\{1, 2, 4\}, \{3, 5\})$, $H(\{1, 2, 5\}, \{3, 4\})$, $H(\{1, 2\}, \{3, 4, 5\})$
4	$H(\{1, 2, 3, 4\})$, $H(\{1, 2, 3, 5\})$, $H(\{1, 2, 4, 5\})$, $H(\{1, 2\}, \{3, 4\})$, $H(\{1, 2\}, \{3, 5\})$, $H(\{1, 2\}, \{4, 5\})$
3	$H(\{1, 2, 3\})$, $H(\{1, 2, 4\})$, $H(\{1, 2, 5\})$
2	$H(\{1, 2\})$

$H(\{1, 2, 3, 4, 5\}) = H_0, \ J = 1, \ \ell_1 = 5$
$H(\{1, 2, 5\}, \{3, 4\}) : \ \mu_1 = \mu_2 = \mu_5, \ \mu_3 = \mu_4, \ J = 2, \ \ell_1 = 3, \ \ell_2 = 2$
$H(\{1, 2\}, \{3, 5\}) : \ \mu_1 = \mu_2, \ \mu_3 = \mu_5, \ J = 2, \ \ell_1 = 2, \ \ell_2 = 2$
$H(\{1, 2, 5\}) : \ \mu_1 = \mu_2 = \mu_5, \ J = 1, \ \ell_1 = 3$
$H(\{1, 2\}) = H_{(1,2)} : \ \mu_1 = \mu_2, \ J = 1, \ \ell_1 = 2$

Tables 1.3 and 1.4, respectively. We limited attention to $m = 60$ and $2 \le M \le 10$. $\ell = M - 1$ is not used in the procedure [1.3]. When $\ell = M = k$ is satisfied, $ta(\ell, m; \alpha(M, \ell)) = ta(k, m; \alpha)$ holds.

When $\alpha = 0.05, \ 0.01$, $m = 60$, $4 \le k \le 10$, from Tables 1.3 and 1.4, we find

$$ta(\ell, m; \alpha(M, \ell)) < ta(k, m; \alpha(k, k)) = ta(k, m; \alpha) \tag{1.26}$$

for ℓ such that $2 \le \ell < M \le k$. By numerical calculation, we verify that (1.26) holds for $50 \le m \le 150$, $\alpha = 0.05, \ 0.01$, and $3 \le k \le 10$. From the construction of the closed testing procedure [1.3] and the relation of (1.26), we get the following (i) and (ii). (i) The procedure [1.3] of level α rejects $H_{(i,i')}$ that is rejected by the Tukey–Kramer simultaneous test [1.1] of level α. (ii) The Tukey–Kramer simultaneous test [1.1] of level α does not always reject $H_{(i,i')}$ that is rejected by the procedure [1.3] of level α. Hence, for $\alpha = 0.05, \ 0.01, 3 \le k \le 10$ and $50 \le m \le 150$, the closed testing procedure [1.3] is more powerful than the single-step Tukey–Kramer simultaneous test [1.1].

We get Lemma 1.1.

Lemma 1.1 *Let $A_{(i,i')}$ be the event that $H_{(i,i')}$ is rejected by the procedure [1.3] as a multiple comparison of level α ($(i, i') \in \mathscr{U}$). Suppose that*

$$ta\,(\ell, m; \alpha(M, \ell)) < ta(k, m; \alpha) \tag{1.27}$$

is satisfied for any M such that $4 \le M \le k$ and any integer ℓ such that $2 \le \ell \le M - 2$, where M is defined by (1.18). Then the following relations hold:

$$\bigcup_{(i,i') \in \mathscr{U}} A_{(i,i')} = \left\{ \max_{(i,i') \in \mathscr{U}} |T_{i'i}| > ta(k, m; \alpha) \right\}, \tag{1.28}$$

$$A_{(i,i')} \supset \{|T_{i'i}| > ta(k, m; \alpha)\} \qquad \left((i, i') \in \mathscr{U}\right). \tag{1.29}$$

Table 1.3 Critical values $ta(\ell, m; \alpha(M, \ell))$ for the stepwise procedure [1.3] with $\alpha = 0.05$ and $m = 60$

$M \setminus \ell$	2	3	4	5	6	7	8	9	10
10	2.653	2.873	2.994	3.075	3.136	3.184	3.224	◊	3.285
9	2.613	2.834	2.955	3.036	3.097	3.145	◊	3.218	
8	2.568	2.790	2.911	2.993	3.053	◊	3.140		
7	2.516	2.740	2.861	2.942	◊	3.051			
6	2.456	2.680	2.802	◊	2.944				
5	2.384	2.609	◊	2.812					
4	2.294	◊	2.643						
3	◊	2.403							
2	2.000								

The places of ◊ are not used in the procedure [1.3]

Proof For $(i, i') \in \mathscr{U}$, we put

$$B_{(i,i')} = \{|T_{i'i}| > ta(k, m; \alpha)\}. \tag{1.30}$$

If $|T_{i'i}| > ta(k, m; \alpha)$ is satisfied, by using (1.27), the procedure [1.3] of level α rejects $H_{(i,i')}$. This implies $B_{(i,i')} \subset A_{(i,i')}$. Hence, we get (1.29) and

$$\bigcup_{(i,i')\in\mathscr{U}} B_{(i,i')} \subset \bigcup_{(i,i')\in\mathscr{U}} A_{(i,i')}. \tag{1.31}$$

If $H_{(i,i')}$ is rejected by the procedure [1.3] of level α, $\max_{(i,i')\in\mathscr{U}} |T_{i'i}| > ta(k, m; \alpha)$ holds. Therefore, we have

$$A_{(i,i')} \subset \bigcup_{(i,i')\in\mathscr{U}} B_{(i,i')}. \tag{1.32}$$

From (1.31) and (1.32), we get

$$\bigcup_{(i,i')\in\mathscr{U}} A_{(i,i')} = \bigcup_{(i,i')\in\mathscr{U}} B_{(i,i')},$$

which implies (1.28). □

From a straightforward application of Lemma 1.1, we get Theorem 1.3.

Theorem 1.3 *Under the assumptions of Lemma 1.1, the following relations hold:*

$$P\left(\bigcup_{(i,i')\in\mathscr{U}} A_{(i,i')}\right) = P\left(\max_{(i,i')\in\mathscr{U}} |T_{i'i}| > ta(k, m; \alpha)\right), \tag{1.33}$$

$$P\left(A_{(i,i')}\right) \geq P\left(|T_{i'i}| > ta(k, m; \alpha)\right) \qquad \left((i, i') \in \mathscr{U}\right). \tag{1.34}$$

Table 1.4 Critical values $ta(\ell, m; \alpha(M, \ell))$ for the stepwise procedure [1.3] with $\alpha = 0.01$ and $m = 60$

$M \setminus \ell$	2	3	4	5	6	7	8	9	10
10	3.230	3.440	3.558	3.638	3.699	3.747	3.787	◊	3.851
9	3.195	3.406	3.523	3.603	3.664	3.713	◊	3.787	
8	3.155	3.366	3.484	3.565	3.625	◊	3.714		
7	3.109	3.322	3.440	3.520	◊	3.630			
6	3.056	3.269	3.388	◊	3.529				
5	2.993	3.207	◊	3.407					
4	2.914	◊	3.249						
3	◊	3.028							
2	2.660								

The left-hand side (l.h.s) of (1.33) is the probability that the procedure [1.3] rejects at least one of the hypotheses in $\mathscr{H}$. The right-hand side (r.h.s.) of (1.33) is the probability that the Tukey–Kramer test rejects at least one of these hypotheses. The l.h.s. of (1.34) is the probability that the procedure [1.3] rejects $H_{(i,i')}$. The r.h.s. of (1.34) is the probability that the Tukey–Kramer test rejects $H_{(i,i')}$. The relation (1.34) means that the l.h.s. is greater than or equal to the r.h.s. For any $\boldsymbol{\mu}$, (1.33) and (1.34) hold.

We consider the $100(1-\alpha)\%$ confidence region for $\boldsymbol{\mu} = (\mu_1, \ldots, \mu_k)$ which is induced by the test procedure [1.3]. Suppose that the condition of Lemma 1.1 is satisfied. We set $\boldsymbol{X} = (X_{11}, \ldots, X_{1n_1}, \ldots, X_{k1}, \ldots, X_{kn_k})$. For any $(i, i') \in \mathscr{U}$, there exists $A^*_{(i,i')}$ such that $A_{(i,i')} = \left\{\boldsymbol{X} \in A^*_{(i,i')}\right\}$. From (1.28), we find

$$\bigcup_{(i,i') \in \mathscr{U}} \left\{\boldsymbol{X} \in A^*_{(i,i')}\right\} = \left\{\max_{(i,i') \in \mathscr{U}} |T_{i'i}| > ta(k, m; \alpha)\right\}. \tag{1.35}$$

The direct product of $\boldsymbol{\mu}$ and $\mathbf{1}_n$ is denoted by $\boldsymbol{\mu} \otimes \mathbf{1}_n = (\mu_1 \mathbf{1}_{n_1}, \ldots, \mu_k \mathbf{1}_{n_k})$, where $\mathbf{1}_{n_i}$ is the row vector consisting of n_i ones. By replacing $\boldsymbol{X}$ with $\boldsymbol{X} - \boldsymbol{\mu} \otimes \mathbf{1}_n$ in (1.35), we have

$$\bigcup_{(i,i') \in \mathscr{U}} \left\{\boldsymbol{X} - \boldsymbol{\mu} \otimes \mathbf{1}_n \in A^*_{(i,i')}\right\} = \left\{\max_{(i,i') \in \mathscr{U}} |T_{i'i}(\boldsymbol{\mu})| > ta(k, m; \alpha)\right\}. \tag{1.36}$$

From (1.36) and (1.33), we get

$$P_{\mu}\left(\bigcap_{(i,i')\in\mathscr{U}}\left\{X-\mu\otimes\mathbf{1}_n\in\left(A^*_{(i,i')}\right)^c\right\}\right)=P_{\mu}\left(\max_{(i,i')\in\mathscr{U}}|T_{i'i}(\mu)|\le ta(k,m;\alpha)\right)$$
$$=P_0\left(\max_{(i,i')\in\mathscr{U}}|T_{i'i}|\le ta(k,m;\alpha)\right)$$
$$=1-\alpha. \tag{1.37}$$

Hence, (1.36) and (1.37) imply that the $100(1-\alpha)\%$ confidence region for $\mu=(\mu_1,\ldots,\mu_k)$ induced by the test procedure [1.3] becomes

$$\bigcap_{(i,i')\in\mathscr{U}}\left\{\mu\mid X-\mu\otimes\mathbf{1}_n\in\left(A^*_{(i,i')}\right)^c\right\}=\left\{\mu\mid\max_{1\le i<i'\le k}|T_{i'i}(\mu)|\le ta(k,m;\alpha)\right\}. \tag{1.38}$$

Since the r.h.s. of (1.38) is equal to (1.13), the $100(1-\alpha)\%$ confidence region for $\mu=(\mu_1,\ldots,\mu_k)$ induced by the test procedure [1.3] is equivalent to $100(1-\alpha)\%$ Tukey–Kramer simultaneous confidence intervals of (1.13).

As a closed testing procedure under assuming normality for k-sample model, the REGW (Ryan-Einot-Gabriel-Welsch) method is utilized. The REGW method is also stated in Hsu (1996). In order to introduce the REGW method, we define the hypothesis $H(I)$ by $H(I):\ \mu_i=\mu_{i'}$ for $i,i'\in I$ and we put $\iota=\#(I)$, where I $(I\subset\{1,\ldots,k\})$ and $\#(I)\ge 2$. Suppose $k\ge 4$. We define $\alpha^*(\iota)$ by

$$\alpha^*(\iota)=\begin{cases}1-(1-\alpha)^{\iota/k} & (2\le\iota\le k-2)\\ \alpha & (\iota=k-1,k).\end{cases} \tag{1.39}$$

[1.4] REGW Method

If $ta\,(\iota,m;\alpha^*(\iota))<T(I)$ for any I such that $i,i'\in I$, $H_{(i,i')}$ is rejected.

Suppose $\ell_j=\iota=\ell$. Then, since

$$1-(1-\alpha)^{\ell/M}\ge 1-(1-\alpha)^{\ell/k},$$

in testing the null hypothesis $\bigwedge_{v\in V}H_v$, the rejection region for the closed testing procedure [1.3] includes the one for the closed testing procedure [1.4]. Therefore, the closed testing procedure [1.3] is more powerful than the closed testing procedure [1.4].

[1.5] Stepwise Procedure Based on F-Statistics

Let us put

$$S(I_j)=\sum_{i\in I_j}n_i\left(\bar{X}_{i\cdot}-\bar{X}_{I_j}\right)^2/\{(\ell_j-1)V_E\}\ (j=1,\ldots,J), \tag{1.40}$$

where I_j is defined in (1.17), ℓ_j is defined in (1.18), and $\bar{X}_{I_j}\equiv\sum_{i\in I_j}n_i\bar{X}_{i\cdot}/\sum_{i\in I_j}n_i$. In the procedure [1.3], replace $ta\left(\ell_j,m;\alpha(M,\ell_j)\right)<T(I_j)$ and $ta\,(M,m;\alpha)<$

$T(I_1)$ with $F_m^{\ell_j - 1}\left(\alpha(M, \ell_j)\right) < S(I_j)$ and $F_m^{M-1}(\alpha) < S(I_1)$, respectively. Then, this procedure also becomes a closed test.

References

Enderton HB (2001) Mathematical introduction to logic, 2nd edn. Academic Press, Cambridge

Hayter AJ (1984) A proof of the conjecture that the Tukey-Kramer multiple comparisons procedure is conservative. Ann Statist 12:61–75

Hsu JC (1996) Multiple comparisons-theory and methods. Chapman and Hall, London

Kramer CY (1956) Extension of multiple range tests to group means with unequal numbers of replications. Biometrics 12:307–310

Shiraishi T (2006) The upper bound for the distribution of Tukey-Kramer's statistic. Bull Comput Statist 19:77–87

Shiraishi T (2011) Closed testing procedures for pairwise comparisons in multi-sample models. Biom Soc Jpn 32:33–47 (in Japanese)

Tukey JW (1953) The problem of multiple comparisons. The collected works of John W. Tukey, (1994) volume viii: multiple comparisons. Chapman and Hall, London

Chapter 2
Multiple Comparisons in Heteroscedastic Multi-sample Models

Abstract We consider multiple comparison procedures among mean effects in heteroscedastic k-sample models. We propose closed testing procedures based on the maximum values of some two-sample Welch t-test statistics. The results reveal that the proposed procedures are more powerful than single-step procedures of Games and Howell (J Educ Stat 1:113–125, 1976).

2.1 Introduction

We consider heteroscedastic k-sample models under normality. $(X_{i1}, \ldots, X_{in_i})$ is a random sample of size n_i from the ith normal population with mean μ_i and variance σ_i^2 $(i = 1, \ldots, k)$, that is, $P(X_{ij} \leq x) = \Phi((x - \mu_i)/\sigma_i)$. Furthermore, X_{ij}'s are assumed to be independent. Then unbiased estimators of μ_i and σ_i^2 $(i = 1, \ldots, k)$ are, respectively, given by $\hat{\mu}_i = \bar{X}_{i\cdot}$ and

$$\tilde{\sigma}_i^2 = \frac{1}{n_i - 1} \sum_{j=1}^{n_i} (X_{ij} - \bar{X}_{i\cdot})^2 \quad (i = 1, \ldots, k). \tag{2.1}$$

We consider test procedures for all-pairwise comparisons of

$$\left\{ \text{the null hypothesis } H_{(i,i')} \text{ vs. the alternative } H^A_{(i,i')} \,\middle|\, (i, i') \in \mathscr{U} \right\},$$

where $H_{(i,i')}$ and $H^A_{(i,i')}$ are defined in (1.4), and $\mathscr{U}$ is defined by (1.5).

Games and Howell (1976) proposed single-step procedures as multiple comparison tests of level α. Shiraishi and Hayakawa (2014) proposed closed testing procedures. It is shown that the proposed multistep procedures are more powerful than the single-step procedures.

T.-a. Shiraishi et al., *Pairwise Multiple Comparisons*,
JSS Research Series in Statistics,
https://doi.org/10.1007/978-981-15-0066-4_2

2.2 The Games–Howell Method

We put

$$T^G_{i'i} := \frac{\bar{X}_{i'\cdot} - \bar{X}_{i\cdot}}{\sqrt{\frac{\tilde{\sigma}_i^2}{n_i} + \frac{\tilde{\sigma}_{i'}^2}{n_{i'}}}}.$$

Furthermore, for $\boldsymbol{\mu} := (\mu_1, \ldots, \mu_k)$, we put

$$T^G_{i'i}(\boldsymbol{\mu}) := \frac{\bar{X}_{i'\cdot} - \bar{X}_{i\cdot} - (\mu_{i'} - \mu_i)}{\sqrt{\frac{\tilde{\sigma}_i^2}{n_i} + \frac{\tilde{\sigma}_{i'}^2}{n_{i'}}}}.$$

Then for $t \geq 0$,

$$P\left(\max_{(i,i')\in\mathscr{U}} |T^G_{i'i}(\boldsymbol{\mu})| \leq t\right) = P_0\left(\max_{(i,i')\in\mathscr{U}} |T^G_{i'i}| < t\right)$$

holds. Corresponding to (1.10), for ℓ such that $2 \leq \ell \leq k$, we put

$$ta(\ell, m_0; \alpha) := \text{a solution of } t \text{ satisfying the equation } TA(t|\ell, m_0) = 1 - \alpha, \tag{2.2}$$

where $TA(t|\ell, m_0)$ is defined by (1.20). Then we can state the following single-step procedures of Games and Howell (1976).

[2.1] Simultaneous Tests
The simultaneous test of level α for the null hypotheses $\{H_{(i,i')}|\ (i, i') \in \mathscr{U}\}$ consists in rejecting $H_{(i,i')}$ for $(i, i') \in \mathscr{U}$ such that $|T^G_{i'i}| > ta(k, \widehat{m}_{i'i}; \alpha)$, where $\widehat{m}_{i'i} := \left[\dfrac{\left(\dfrac{\tilde{\sigma}_i^2}{n_i} + \dfrac{\tilde{\sigma}_{i'}^2}{n_{i'}}\right)^2}{\dfrac{\tilde{\sigma}_i^4}{n_i^2(n_i - 1)} + \dfrac{\tilde{\sigma}_{i'}^4}{n_{i'}^2(n_{i'} - 1)}} + \dfrac{1}{2}\right]$ and $[x]$ denotes the maximum integer that does not exceed x.

[2.2] Simultaneous Confidence Intervals
$100(1 - \alpha)\%$ Games–Howell simultaneous confidence intervals for all-pairwise $\{\mu_{i'} - \mu_i|\ (i, i') \in \mathscr{U}\}$ are given by

$$\mu_{i'} - \mu_i \in \bar{X}_{i'\cdot} - \bar{X}_{i\cdot} \pm ta(k, \widehat{m}_{i'i}; \alpha) \cdot \sqrt{\frac{\tilde{\sigma}_i^2}{n_i} + \frac{\tilde{\sigma}_{i'}^2}{n_{i'}}} \quad ((i, i') \in \mathscr{U}). \tag{2.3}$$

2.3 Closed Testing Procedures

We consider closed testing procedures. For an integer J and disjoint sets $I_1, \ldots, I_J \subset \{1, \ldots, k\}$, we define the null hypothesis $H(I_1, \ldots, I_J)$ by (1.16). For any nonempty $V \subset \mathscr{U}$, there exist an integer J and disjoint sets $I_1, \ldots, I_J$ such that

$$\bigwedge_{v \in V} H_v = H(I_1, \ldots, I_J) \tag{2.4}$$

and $\#(I_j) \geq 2$ $(j = 1, \ldots, J)$. For $H(I_1, \ldots, I_J)$ of (2.4), we define M and ℓ_j by (1.18).

[2.3] Stepwise Procedure Based on Welch's Statistics
For $\ell = \ell_1, \ldots, \ell_J$, we define $\alpha(M, \ell)$ by (1.19).

(a) $J \geq 2$
Whenever $ta(\ell_j, \widehat{m}_{i'i}; \alpha(M, \ell_j)) \leq |T_{i'i}^G|$ holds for some integers j and $i < i'$ such that $1 \leq j \leq J$ and $i, i' \in I_j$, we reject the hypothesis $\bigwedge_{v \in V} H_v$.

(b) $J = 1$ $(M = \ell_1)$
Whenever $ta(M, \widehat{m}_{i'i}; \alpha) \leq |T_{i'i}^G|$ holds for some integers $i < i'$ such that $i, i' \in I_j$, we reject the hypothesis $\bigwedge_{v \in V} H_v$.

By using the methods of (a) and (b), when $\bigwedge_{v \in V} H_v$ is rejected for any V such that $(i, i') \in V \subset \mathscr{U}$, the null hypothesis $H_{(i,i')}$ is rejected as a multiple comparison test.

We get the following theorem by Shiraishi and Hayakawa (2014).

Theorem 2.1 *The test procedure [2.3] is a multiple comparison test of level* α.

We consider $k = 4$. From Table 2.1, whenever the following (2-a1)–(2-a5) are satisfied, the closed testing procedure of [2.3] rejects $H_{(1,2)}$ as a multiple comparison test of level α.

(2-a1) $|T_{i'i}^G| > ta(4, \widehat{m}_{i'i}; \alpha)$ for some integers i, i' such that $1 \leq i < i' \leq 4$.
(2-a2) $|T_{21}^G| > ta(2, \widehat{m}_{21}; \alpha(4, 2))$ or $|T_{43}^G| > ta(2, \widehat{m}_{43}; \alpha(4, 2))$.
(2-a3) $|T_{i'i}^G| > ta(3, \widehat{m}_{i'i}; \alpha)$ for some integers i, i' such that $1 \leq i < i' \leq 3$.
(2-a4) $|T_{21}^G| > ta(3, \widehat{m}_{21}; \alpha)$ or $|T_{41}^G| > ta(3, \widehat{m}_{41}; \alpha)$ or $|T_{42}^G| > ta(3, \widehat{m}_{42}; \alpha)$.
(2-a5) $|T_{21}^G| > ta(2, \widehat{m}_{21}; \alpha)$.

2.4 Asymptotic Theory

In order to discuss the asymptotic theory of the Games–Howell methods and the closed testing procedure [2.3], we add the condition (C1).

$$\textbf{(C1)} \qquad \lim_{n\to\infty}(n_i/n) = \lambda_i > 0 \ \ (1 \le i \le k) \tag{2.5}$$

Then we get Lemma 2.1.

Lemma 2.1 *Under the condition (C1), we have, as n tends to infinity,*

$$\max_{(i,i')\in\mathscr{U}} \left|T^G_{i'i}(\boldsymbol{\mu})\right| \xrightarrow{\mathscr{L}} \max_{(i,i')\in\mathscr{U}} \frac{|Y_{i'} - Y_i|}{\sqrt{\frac{\sigma_i^2}{\lambda_i} + \frac{\sigma_{i'}^2}{\lambda_{i'}}}},$$

where $Y_1, \ldots, Y_k$ are independent and $Y_i \sim N(0, \sigma_i^2/\lambda_i)$ $(i = 1, \ldots, k)$.

Proof From a central limit theorem, we get

$$\sqrt{n_i}(\bar{X}_{i\cdot} - \mu_i) \xrightarrow{\mathscr{L}} \sigma_i Z_i \sim N(0, \sigma_i^2),$$

where $Z_i \sim N(0, 1)$, $Z_1, \ldots, Z_k$ are independent and identically distributed with the standard normal distribution. It follows that

$$\sqrt{n}(\bar{X}_{i\cdot} - \mu_i) = \left(\frac{\sqrt{n}}{\sqrt{n_i}}\right)\sqrt{n_i}(\bar{X}_{i\cdot} - \mu_i) \xrightarrow{\mathscr{L}} \frac{\sigma_i}{\sqrt{\lambda_i}} Z_i \sim N\left(0, \frac{\sigma_i^2}{\lambda_i}\right).$$

By using Khintchin's law of large numbers, we have $\tilde{\sigma}_i^2 \xrightarrow{P} \sigma_i^2$. From Slutsky's theorem, it follows that

$$T^G_{i'i}(\boldsymbol{\mu}) = \frac{\sqrt{n}(\bar{X}_{i'\cdot} - \mu_{i'}) - \sqrt{n}(\bar{X}_{i\cdot} - \mu_i)}{\sqrt{\frac{n}{n_i}\tilde{\sigma}_i^2 + \frac{n}{n_{i'}}\tilde{\sigma}_{i'}^2}} \xrightarrow{\mathscr{L}} \frac{\frac{\sigma_{i'}}{\sqrt{\lambda_{i'}}} Z_{i'} - \frac{\sigma_i}{\sqrt{\lambda_i}} Z_i}{\sqrt{\frac{\sigma_i^2}{\lambda_i} + \frac{\sigma_{i'}^2}{\lambda_{i'}}}} \sim N(0, 1).$$

Therefore, we get the assertion

$$\max_{(i,i')\in\mathscr{U}} \left|T^G_{i'i}(\boldsymbol{\mu})\right| \xrightarrow{\mathscr{L}} \max_{(i,i')\in\mathscr{U}} \frac{\left|\frac{\sigma_{i'}}{\sqrt{\lambda_{i'}}} Z_{i'} - \frac{\sigma_i}{\sqrt{\lambda_i}} Z_i\right|}{\sqrt{\frac{\sigma_i^2}{\lambda_i} + \frac{\sigma_{i'}^2}{\lambda_{i'}}}}$$

$$\Longleftrightarrow \max_{(i,i')\in\mathscr{U}} \left|T^G_{i'i}(\boldsymbol{\mu})\right| \xrightarrow{\mathscr{L}} \max_{(i,i')\in\mathscr{U}} \frac{|Y_{i'} - Y_i|}{\sqrt{\frac{\sigma_i^2}{\lambda_i} + \frac{\sigma_{i'}^2}{\lambda_{i'}}}}. \tag{2.6}$$

□

In Lemma 2.1, by replacing the condition $X_{ij} \sim N(\mu_i, \sigma_i^2)$ with the general condition (C2), the assertion of Lemma 2.1 retains to hold.

(C2) $P(X_{ij} \le x) = F\left(\frac{x-\mu_i}{\sigma_i}\right)$, $\int_{-\infty}^{\infty} x dF(x)dx = 0$, $\int_{-\infty}^{\infty} x^2 dF(x)dx = 1$,
where $F(x)$ is a continuous distribution function.

From Lemma 2.1, we get Proposition 2.1.

Proposition 2.1 *Under the condition of Lemma 2.1, we have*

$$A(t) \le \lim_{n\to\infty} P\left(\max_{(i,i')\in\mathscr{U}} |T_{i'i}^G(\boldsymbol{\mu})| \le t\right) \le A_\sigma^*(t), \tag{2.7}$$

where

$$A(t) :=k\int_{-\infty}^{\infty}\{\Phi(x)-\Phi(x-\sqrt{2}\cdot u)\}^{k-1}d\Phi(x), \tag{2.8}$$

$$A_\sigma^*(t) := \int_{-\infty}^{\infty}\sum_{i'=1}^{k}\prod_{\substack{i=1\\ i\ne i'}}^{k}\left\{\Phi\left(\sqrt{\frac{\lambda_i\sigma_{i'}^2}{\lambda_{i'}\sigma_i^2}}\cdot x\right)\right.$$
$$\left. - \Phi\left(\sqrt{\frac{\lambda_i\sigma_{i'}^2}{\lambda_{i'}\sigma_i^2}}\cdot x - \sqrt{\frac{\lambda_i\sigma_{i'}^2+\lambda_{i'}\sigma_i^2}{\lambda_{i'}\sigma_i^2}}\cdot t\right)\right\} d\Phi(x).$$

When $\sigma_1^2/\lambda_1 = \cdots = \sigma_k^2/\lambda_k$ is satisfied, both of the inequalities of (2.7) become an equality.

Proof From Lemma 2.1, for $t \ge 0$,

$$\lim_{n\to\infty} P\left(\max_{(i,i')\in\mathscr{U}} \left|T_{i'i}^G(\boldsymbol{\mu})\right| \le t\right) = P\left(\max_{(i,i')\in\mathscr{U}} \frac{|Y_i - Y_{i'}|}{\sqrt{\frac{\sigma_i^2}{\lambda_i}+\frac{\sigma_{i'}^2}{\lambda_{i'}}}} \le t\right) \tag{2.9}$$

holds, where $Y_i \sim N(0, \sigma_i^2/\lambda_i)$. By replacing σ_i^2 with σ_i^2/λ_i in Theorem A.5 of Shiraishi (2011), we derive (2.7). □

For a given α such that $0 < \alpha < 1$, we put

$$a(k;\alpha) := \text{a solution of } t \text{ satisfying the equation } A(t) = 1-\alpha. \tag{2.10}$$

When $n_1, \ldots, n_k$ are large, from Proposition 2.1, we get asymptotic Games–Howell methods.

[2.4] Asymptotic Games–Howell Methods
The asymptotic simultaneous test of level α for the null hypotheses $\{H_{(i,i')} | (i,i') \in \mathscr{U}\}$ consists in rejecting $H_{(i,i')}$ for $(i,i') \in \mathscr{U}$ such that $|T_{i'i}^G| > a(k;\alpha)$. From (2.9), we find that $100(1-\alpha)\%$ asymptotic simultaneous confidence intervals for all-pairwise $\{\mu_{i'} - \mu_i | (i,i') \in \mathscr{U}\}$ are given by

Table 2.1 The values of $a(k;\alpha)$ for $\alpha = 0.05,\ 0.01$, and $k = 2(1)10$

$A(a(k;\alpha)) = 1-\alpha \rightarrow a(k;\alpha)$

$100\alpha\%\backslash k$	2	3	4	5	6	7	8	9	10
5%	1.960	2.344	2.569	2.728	2.850	2.948	3.031	3.102	3.164
1%	2.576	2.913	3.113	3.255	3.364	3.452	3.526	3.590	3.646

Table 2.2 The values of $a(\ell;\alpha(M,\ell))$ for $\alpha = 0.05$

$M\setminus\ell$	2	3	4	5	6	7	8	9	10
10	2.569	2.774	2.887	2.964	3.021	3.066	3.104	◊	3.164
9	2.532	2.739	2.852	2.929	2.986	3.032	◊	3.102	
8	2.491	2.699	2.813	2.890	2.947	◊	3.031		
7	2.443	2.653	2.767	2.845	◊	2.948			
6	2.388	2.599	2.714	◊	2.850				
5	2.321	2.534	◊	2.728					
4	2.236	◊	2.569						
3	◊	2.344							
2	1.960								

The places of ◊ are not used in the procedure [2.5]

$$\mu_{i'} - \mu_i \in \bar{X}_{i'\cdot} - \bar{X}_{i\cdot} \pm a(k;\alpha)\cdot\sqrt{\frac{\tilde{\sigma}_i^2}{n_i} + \frac{\tilde{\sigma}_{i'}^2}{n_{i'}}} \quad ((i,i') \in \mathscr{U}).$$

Let us put

$$T^G(I_j) = \max_{i<i',\ i,i'\in I_j} \left|T_{i'i}^G\right| \quad (j = 1,\ldots,J).$$

Then, we propose the stepwise procedure [2.5].

[2.5] Asymptotic Closed Testing Procedures

For $\ell = \ell_1,\ldots,\ell_J$, we define $\alpha(M,\ell)$ by (1.19). Corresponding to (2.8), we put

$$A(t|\ell) = \ell\int_{-\infty}^{\infty}\{\Phi(x) - \Phi(x-\sqrt{2}\cdot ts)\}^{\ell-1}d\Phi(x). \tag{2.11}$$

By obeying the notation $a(k;\alpha)$, we get

$$A(a(\ell;\alpha(M,\ell))|\ell) = 1-\alpha(M,\ell) = (1-\alpha)^{\ell/M},$$

that is, $a(\ell;\alpha(M,\ell))$ is an upper $100\alpha(M,\ell)\%$ point of the distribution $A(t|\ell)$.

(a) $J \geq 2$
Whenever $a\left(\ell_j;\alpha(M,\ell_j)\right) < T^G(I_j)$ holds for an integer j such that $1 \leq j \leq J$, we reject the hypothesis $\bigwedge_{v\in V} H_v$.

Table 2.3 The values of $a\ (\ell; \alpha(M, \ell))$ for $\alpha = 0.01$

$M \setminus \ell$	2	3	4	5	6	7	8	9	10
10	3.089	3.277	3.382	3.454	3.508	3.552	3.588	◊	3.646
9	3.058	3.247	3.352	3.424	3.479	3.523	◊	3.590	
8	3.022	3.213	3.318	3.391	3.446	◊	3.526		
7	2.982	3.173	3.280	3.353	◊	3.452			
6	2.934	3.128	3.235	◊	3.364				
5	2.877	3.073	◊	3.255					
4	2.806	◊	3.113						
3	◊	2.913							
2	2.576								

(b) $J = 1\ (M = \ell_1)$
Whenever $a\ (M; \alpha) < T^G(I_1)$, we reject the hypothesis $\bigwedge_{v \in V} H_v$.

By using the methods of (a) and (b), when $\bigwedge_{v \in V} H_v$ is rejected for any V such that $(i, i') \in V \subset \mathscr{U}$, the null hypothesis $H_{(i,i')}$ is rejected as an asymptotic multiple comparison test.

We get the following theorem by Shiraishi and Hayakawa (2014).

Theorem 2.2 *Under the condition (C1) of (2.5), as n tends to infinity, the test procedure [2.5] is an asymptotic multiple comparison test of level* α*.*

We consider $k = 4$. From Table 2.1, whenever the following (2-b1)–(2-b5) are satisfied, the closed testing procedure of [2.5] rejects $H_{(1,2)}$ as a multiple comparison test of level α. For $\alpha = 0.05,\ 0.01$, we give the values of $a(\ell; \alpha(M, \ell))$ in Tables 2.2 and 2.3, respectively.

(2-b1) $T^G(\{1, 2, 3, 4\}) = \max_{1 \le i < i' \le 5} |T^G_{i'i}| > a(4; \alpha)$.
(2-b2) $T^G(\{1, 2\}) = |T^G_{21}| > a(2; \alpha(4, 2))$ or $T^G(\{3, 4\}) = |T^G_{43}| > a(2; \alpha(4, 2))$.
(2-b3) $T^G(\{1, 2, 3\}) = \max_{1 \le i < i' \le 3} |T^G_{i'i}| > a(3; \alpha)$.
(2-b4) $T^G(\{1, 2, 4\}) = \max\{|T^G_{21}|, |T^G_{41}|, |T^G_{42}|\} > a(3; \alpha)$.
(2-b5) $T^G(\{1, 2\}) = |T^G_{21}| > a(2; \alpha)$.

References

Games PA, Howell JF (1976) Pairwise multiple comparison procedures with unequal N's and/or variances: a Monte Carlo study. J Educ Stat 1:113–125

Shiraishi T (2011) Multiple comparison procedures under continuous distributions. Kyoritsu-Shuppan Co., Ltd., Tokyo (in Japanese)

Shiraishi T, Hayakawa Y (2014) Closed testing procedures for pairwise multiple comparisons in multi-sample models with unequal variances. Biom Soc Jpn 35:55–68 (in Japanese)

Chapter 3
Multiple Comparison Procedures Under Simple Order Restrictions

Abstract Under the assumptions of continuous distribution and homogeneous variance, we consider multiple comparisons tests for the differences among mean responses in k-samples. Hayter (J Am Stat Assoc 85:778–785, 1990) proposed a single-step procedure as all-pairwise comparisons between ordered means in k-samples under normality. We propose closed testing procedures which are superior to the single-step procedure.

3.1 Introduction

We consider homoscedastic k-sample models under normality. $(X_{i1}, \ldots, X_{in_i})$ is a random sample of size n_i from the ith normal population with mean μ_i and variance σ^2 $(i = 1, \ldots, k)$, that is, $P(X_{ij} \leq x) = \Phi((x - \mu_i)/\sigma)$. Furthermore, X_{ij}'s are assumed to be independent.

When the simple order restrictions

$$\mu_1 \leq \mu_2 \leq \cdots \leq \mu_k \tag{3.1}$$

is satisfied, we consider the null hypothesis H_0 versus the alternative $H^A : \mu_1 \leq \mu_2 \leq \cdots \leq \mu_k$ with at least one strict inequality, which is equivalent to $H_0 : \mu_1 = \mu_k$ versus $H^A : \mu_1 < \mu_k$, where H_0 is defined by (1.3). We define $\{\hat{\mu}_i^* \mid i = 1, \ldots, k\}$ by $\{u_i \mid i = 1, \ldots, k\}$ which minimize $\sum_{i=1}^k \lambda_{ni} \left(u_i - \bar{X}_{i\cdot}\right)^2$ under simple order restrictions $u_1 \leq u_2 \leq \cdots \leq u_k$, that is,

$$\sum_{i=1}^k \lambda_{ni} \left(\hat{\mu}_i^* - \bar{X}_{i\cdot}\right)^2 = \min_{u_1 \leq \cdots \leq u_k} \sum_{i=1}^k \lambda_{ni} \left(u_i - \bar{X}_{i\cdot}\right)^2,$$

where

$$\lambda_{ni} = n_i/n \ (i = 1, \ldots, k). \tag{3.2}$$

$\hat{\mu}_1^*, \ldots, \hat{\mu}_k^*$ are computed by using the pool-adjacent-violators algorithm stated in Robertson et al. (1988). Accordingly, we find

T.-a. Shiraishi et al., *Pairwise Multiple Comparisons*,
JSS Research Series in Statistics,
https://doi.org/10.1007/978-981-15-0066-4_3

$$\tilde{\mu}_i^* = \max_{1 \le p \le i} \min_{i \le q \le k} \frac{\sum_{j=p}^{q} \lambda_{nj} \bar{X}_{j\cdot}}{\sum_{j=p}^{q} \lambda_{nj}} = \max_{1 \le p \le i} \min_{i \le q \le k} \frac{\sum_{j=p}^{q} n_j \bar{X}_{j\cdot}}{\sum_{j=p}^{q} n_j}. \tag{3.3}$$

We put

$$\bar{\chi}_k^2 := \frac{1}{\sigma^2} \sum_{i=1}^{k} n_i \left(\hat{\mu}_i^* - \sum_{j=1}^{k} \lambda_{nj} \bar{X}_{j\cdot} \right)^2. \tag{3.4}$$

We define $\tilde{\nu}_1^*, \ldots, \tilde{\nu}_k^*$ by

$$\sum_{i=1}^{k} \lambda_{ni} \left(\tilde{\nu}_i^* - Y_i \right)^2 = \min_{u_1 \le \cdots \le u_k} \sum_{i=1}^{k} \lambda_{ni} \left(u_i - Y_i \right)^2,$$

where $Y_1, \ldots, Y_k$ are independent and $Y_i \sim N(0, 1/\lambda_{ni})$ $(i = 1, \ldots, k)$. Let $P(L, k; \boldsymbol{\lambda}_n)$ be the probability that $\tilde{\nu}_1^*, \ldots, \tilde{\nu}_k^*$ takes exactly L distinct values, where $\boldsymbol{\lambda}_n = (\lambda_{n1}, \ldots, \lambda_{nk})$. Then, for positive constant c, $P(L, k; c\boldsymbol{\lambda}_n) = P(L, k; \boldsymbol{\lambda}_n)$ holds. Furthermore, from Theorem 2.3.1 of Robertson et al. (1988), we get

$$P_0(\bar{\chi}_k^2 \ge t) = \sum_{L=2}^{k} P(L, k; \boldsymbol{\lambda}_n) P\left(\chi_{L-1}^2 \ge t \right) \qquad (t > 0), \tag{3.5}$$

where $P_0(\cdot)$ denotes the probability measure under H_0 and χ_{L-1}^2 is a chi-square variable with $L - 1$ degrees of freedom. The recurrence formula of computing $P(L, k; \boldsymbol{\lambda}_n)$ is written in Robertson et al. (1988). $P(L, k; \boldsymbol{\lambda}_n)$ is referred to as level probability. We shall state the recurrence formula along the above mentioned notations. Miwa et al. (2000) showed that the value of

$$P(k, k; \boldsymbol{\lambda}_n) = P\left(Y_1 < Y_2 < \cdots < Y_k \right) \tag{3.6}$$

can be derived by the recurrence of one-dimensional computational integration. Let $I_1^d, I_2^d, \ldots, I_L^d$ be a partition of $\{1, 2, \ldots, k\}$ satisfying the following property (P1).
(P1) Each I_s^d is a nonempty set composed of consecutive integers or an integer. When $L \ge 2$, the maximum value of the elements for I_s^d is less than the minimum value of I_{s+1}^d for any integer s such that $1 \le s \le L - 1$.

Then Theorem 2.4.1 of Robertson et al. (1988) gives, for $L = 2, \ldots, k - 1$,

$$P(L, k; \boldsymbol{\lambda}_n) = \sum_{\{I_1^d, I_2^d, \ldots, I_L^d\}} P\left(L, L; \Lambda(I_1^d), \Lambda(I_2^d), \ldots, \Lambda(I_L^d) \right) \cdot \prod_{s=1}^{L} P(1, \#(I_s^d); \boldsymbol{\lambda}_n(I_s^d)), \tag{3.7}$$

where $\#(I_s^d)$ denotes the number of elements of I_s^d, $\Lambda(I_s^d) = \sum_{i \in I_s^d} \lambda_{ni}$, $\boldsymbol{\lambda}_n(I_s^d) = (\lambda_{ni}, \lambda_{ni+1}, \ldots, \lambda_{nj})$ for $I_s^d = \{i, i+1, \ldots, j\}$, and $\sum_{\{I_1^d, I_2^d, \ldots, I_L^d\}}$ denotes the sum over all partitions of $\{1, 2, \ldots, k\}$ satisfying (P1). $\#(I_s^d)$ of (3.7) is less than or equal to $k-1$. Furthermore, we get

$$P(1, k; \boldsymbol{\lambda}_n) = 1 - \sum_{L=2}^{k} P(L, k : \boldsymbol{\lambda}_n)$$

and

$$P(1, 1; \lambda_{ni}) = 1 \quad (1 \leq i \leq k) \tag{3.8}$$

$$P(1, 2; \lambda_{ni}, \lambda_{nj}) = P(2, 2; \lambda_{ni}, \lambda_{nj}) = \frac{1}{2} \quad (1 \leq i < j \leq k). \tag{3.9}$$

Since $P(L, k; \boldsymbol{\lambda}_n)$ depends on L and k for

$$\lambda_{n1} = \cdots = \lambda_{nk} = 1/k, \tag{3.10}$$

we simply write $P(L, k)$ instead of $P(L, k; \boldsymbol{\lambda}_n)$. Barlow et al. (1972) offer the following recurrence formula:

$$\begin{aligned} P(1, k) &= \frac{1}{k}, \\ P(L, k) &= \frac{1}{k} \{(k-1)P(L, k-1) + P(L-1, k-1)\}, \qquad (2 \leq L \leq k-1) \\ P(k, k) &= \frac{1}{k!}. \end{aligned}$$

In $\bar{\chi}_k^2$ defined by (3.4), replace σ^2 with the estimator V_E. Then the subsequent statistic is denoted by

$$\bar{B}^2 := \frac{\sum_{i=1}^{k} n_i (\tilde{\mu}_i^* - \bar{X}_{..})^2}{V_E} = \frac{\bar{\chi}_k^2}{V_E/\sigma^2}. \tag{3.11}$$

Since $\bar{\chi}_k^2$ and V_E are independent, from (3.5), (3.11), and the relationship with $mV_E/\sigma^2 \sim \chi_m^2$, we find, for $t > 0$,

$$P_0(\bar{B}^2 \geq t) = \sum_{L=2}^{k} P(L, k; \boldsymbol{\lambda}_n) P\left((L-1)F_m^{L-1} \geq t\right), \tag{3.12}$$

where F_m^{L-1} denotes the random variable having the F-distribution with $L-1$ and m degrees of freedom. The relation of (3.12) is stated in Miwa et al. (2000). For a given α such that $0 < \alpha < 1$, we give the following equation of t:

$$\sum_{L=2}^{k} P(L, k; \boldsymbol{\lambda}_n) P\left((L-1)F_m^{L-1} \geq t\right) = \alpha$$

We denote a solution of this equation by $\bar{b}^2(k, \boldsymbol{\lambda}_n, m; \alpha)$. Thus, from (3.12), as a test of level α for the null hypothesis H_0 versus the alternative $H^{OA} : \mu_1 < \mu_k$, we can propose to reject H_0 when the value of $\bar{B}^2$ is greater than $\bar{b}^2(k, \boldsymbol{\lambda}_n, m; \alpha)$.

For a given α such that $0 < \alpha < 1$, we give the following equation of t:

$$\sum_{L=2}^{k} P(L, k; \boldsymbol{\lambda}_n) P\left(\chi^2_{L-1} \geq t\right) = \alpha.$$

We denote a solution of this equation by $\bar{c}_k(\boldsymbol{\lambda}_n; \alpha)$, that is, hence, from (3.5), we can reject H_0 when the value of $\bar{\chi}^2_k$ is greater than $\bar{c}_k(\boldsymbol{\lambda}_n; \alpha)$.

For specified i, i' such that $(i, i') \in \mathscr{U}$, if we are interested in testing of

$$\text{the null hypothesis } H_{(i,i')} : \mu_i = \mu_{i'} \text{ vs. the alternative } H^{OA}_{(i,i')} : \mu_i < \mu_{i'}, \quad (3.13)$$

we can use the one-sided two-sample t-test. In the present chapter, we consider test procedures for all-pairwise comparisons of $\left\{\text{the null hypothesis } H_{(i,i')} \text{ vs. the alternative } H^{OA}_{(i,i')} \,\middle|\, (i, i') \in \mathscr{U}\right\}$, where $\mathscr{U}$ is defined by (1.5). Under the equality of sample sizes $n_1 = \cdots = n_k$, Hayter (1990) proposed single-step simultaneous tests for $\left\{\text{the null hypothesis } H_{(i,i')} \text{ vs. the alternative } H^{OA}_{(i,i')} \,\middle|\, (i, i') \in \mathscr{U}\right\}$. Shiraishi (2014) proposed closed testing procedures. It is shown that (i) the proposed multi-step procedures are more powerful than the single-step procedure of Hayter (1990), and (ii) confidence regions induced by the multistep procedures are equivalent to simultaneous confidence intervals.

3.2 Multiple Comparisons Under Equal Sample Sizes

We add the condition (C3) of equal sample sizes throughout this section.

$$(C3) \qquad n_1 = n_2 = \cdots = n_k. \quad (3.14)$$

Then $T_{i'i}$ of (1.8) and $T_{i'i}(\boldsymbol{\mu})$ of (1.11) are, respectively, given by

$$T_{i'i} = \frac{\sqrt{n_1}(\bar{X}_{i'\cdot} - \bar{X}_{i\cdot})}{\sqrt{2V_E}}, \quad (3.15)$$

$$T_{i'i}(\boldsymbol{\mu}) = \frac{\sqrt{n_1}(\bar{X}_{i'\cdot} - \bar{X}_{i\cdot} - \mu_{i'} + \mu_i)}{\sqrt{2V_E}}.$$

We put

$$D_1(t) := P\left(\max_{1\le i<i'\le k} \frac{Z_{i'} - Z_i}{\sqrt{2}} \le t\right), \tag{3.16}$$

where $Z_i \sim N(0, 1)$ and $Z_1, \ldots, Z_k$ are independent. Shiraishi (2014) gives

$$\lim_{n\to\infty} P_0\left(\max_{1\le i<i'\le k} T_{ii'} \le t\right) = D_1(t). \tag{3.17}$$

Let U_E be a random variable distributed to χ^2-distribution with m degrees of freedom that is independent of $Z_1, \ldots, Z_k$. Then we define $TD_1(t)$ by

$$TD_1(t) := P\left(\max_{1\le i<i'\le k} \frac{Z_{i'} - Z_i}{\sqrt{2U_E/m}} \le t\right) = \int_0^\infty D_1(ts)g(s|m)ds, \tag{3.18}$$

where $g(s|m)$ is defined by (1.7). Here we get

$$P\left(\max_{1\le i<i'\le k} T_{i'i}(\boldsymbol{\mu}) \le t\right) = P_0\left(\max_{1\le i<i'\le k} T_{i'i} \le t\right) = TD_1(t). \tag{3.19}$$

Since $-Z_i$ and Z_i have the same distribution,

$$D_1(t) = P\left(\max_{1\le i<i'\le k} \frac{Z_i - Z_{i'}}{\sqrt{2}} \le t\right)$$

holds. We put

$$H_1(t, x) := P\left(\frac{Z_1 - x}{\sqrt{2}} \le t\right) = \Phi(\sqrt{2}\cdot t + x). \tag{3.20}$$

For $r \ge 2$, we have

$$\begin{aligned} H_r(t, x) &= P\left(\max_{1\le i<i'\le r+1} \frac{Z_i - Z_{i'}}{\sqrt{2}} \le t \,\middle|\, Z_{r+1} = x\right) \\ &= P\left(\max_{1\le i<i'\le r} (Z_i - Z_{i'}) \le \sqrt{2}t,\ \max_{1\le i\le r} (Z_i - Zr + 1) \le \sqrt{2}t \,\middle|\, Z_{r+1} = x\right) \\ &= P\left(\max_{1\le i<i'\le r} (Z_i - Z_{i'}) \le \sqrt{2}t,\ \max_{1\le i\le r} (Z_i - x) \le \sqrt{2}t\right). \end{aligned} \tag{3.21}$$

By using the relation of (3.21), we derive

$$\begin{aligned} H_r(t, x) &= P\left(\max_{1\le i<i'\le r} (Z_i - Z_{i'}) \le \sqrt{2}t,\ \max_{1\le i\le r-1} (Z_i - x) \le \sqrt{2}t,\ (Z_r - x) \le \sqrt{2}t\right) \\ &= \int_{-\infty}^{x+\sqrt{2}t} P\left(\max_{1\le i<i'\le r} (Z_i - Z_{i'}) \le \sqrt{2}t,\right. \\ &\qquad \left. \max_{1\le i\le r-1} (Z_i - x) \le \sqrt{2}t \,\middle|\, Z_r = w\right)\varphi(w)dw \end{aligned}$$

$$= \int_{-\infty}^{x+\sqrt{2}t} P\Big(\max_{1\le i<i'\le r-1} (Z_i - Z_{i'}) \le \sqrt{2}t,$$

$$\max_{1\le i\le r-1} Z_i \le \min\left\{x + \sqrt{2}t,\ w + \sqrt{2}t\right\} \Big)\varphi(w)dw$$

$$= \int_{-\infty}^{x} P\Big(\max_{1\le i<i'\le r-1} (Z_i - Z_{i'}) \le \sqrt{2}t,\ \max_{1\le i\le r-1} Z_i \le w + \sqrt{2}t \Big)\varphi(w)dw$$

$$+ \int_{x}^{x+\sqrt{2}t} P\Big(\max_{1\le i<i'\le r-1} (Z_i - Z_{i'}) \le \sqrt{2}t,\ \max_{1\le i\le r-1} Z_i \le x + \sqrt{2}t \Big)\varphi(w)dw$$

$$= \int_{-\infty}^{x} H_{r-1}(t, w)\varphi(w)dw + H_{r-1}(t, x)\{\Phi(\sqrt{2}\cdot t + x) - \Phi(x)\}.$$

Therefor $H_r(t, x)$ and $D_1(t)$ are, respectively, expressed by

$$H_r(t, x) = \int_{-\infty}^{x} H_{r-1}(t, y)\varphi(y)dy + H_{r-1}(t, x)\{\Phi(\sqrt{2}\cdot t + x) - \Phi(x)\} \quad (2 \le r \le k-1) \tag{3.22}$$

$$D_1(t) = \int_{-\infty}^{\infty} H_{k-1}(t, x)\varphi(x)dx. \tag{3.23}$$

Furthermore, from (3.2) and (3.23), we get

$$TD_1(t) = \int_0^{\infty} \left\{ \int_{-\infty}^{\infty} H_{k-1}(ts, x)\varphi(x)dx \right\} g(s|m)ds. \tag{3.24}$$

For a given α such that $0 < \alpha < 1$, we put

$$td_1(k, m; \alpha) := \text{a solution of } t \text{ satisfying the equation } TD_1(t) = 1 - \alpha. \tag{3.25}$$

By using (3.19), we can derive single-step procedures proposed by Hayter (1990).

[3.1] Single-Step Tests Based on One-Sided t-Test Statistics

The simultaneous test of level α for the {null hypothesis $H_{(i,i')}$ vs. alternative hypothesis $H_{(i,i')}^{OA} : \mu_i < \mu_{i'} \mid (i, i') \in \mathscr{U}$} consists in rejecting $H_{(i,i')}$ for $(i, i') \in \mathscr{U}$ such that $T_{i'i} \ge td_1(k, m; \alpha)$, where $H_{(i,i')}$ is defined in (1.4).

[3.2] Simultaneous Confidence Intervals

$100(1-\alpha)\%$ simultaneous confidence intervals for all-pairwise $\{\mu_{i'} - \mu_i \mid (i, i') \in \mathscr{U}\}$ are given by

$$\bar{X}_{i'\cdot} - \bar{X}_{i\cdot} - td_1(k, m; \alpha)\cdot\sqrt{\frac{2V_E}{n_1}} < \mu_{i'} - \mu_i < \infty \ ((i, i') \in \mathscr{U}). \tag{3.26}$$

Next we introduce closed testing procedures. The closure of $\mathscr{H}$ under the order restrictions (3.1) is given by

$$\overline{\mathscr{H}}^o = \left\{ \bigwedge_{v \in V} H_v \,\middle|\, \emptyset \subsetneq V \subset \mathscr{U} \right\} = \left\{ \bigwedge_{v \in V^+} H_v \,\middle|\, \emptyset \subsetneq V \subset \mathscr{U} \right\}, \tag{3.27}$$

where $\mathscr{H}$ is defined by (1.14) and

$$V^+ := \{(i, i+1) | \text{ For } (i_0, i_0') \in V,\ i_0 \leq i < i+1 \leq i_0'\}. \tag{3.28}$$

Then, we get

$$\bigwedge_{v \in V^+} H_v : \text{for any } (i, i') \in V,\ \mu_i = \mu_{i'} \text{ holds}. \tag{3.29}$$

Let $I_1, \ldots, I_J$ be disjoint sets satisfying the following property (C4).

(C4) There exist integers $\ell_1, \ldots, \ell_J \geq 2$ and integers $0 \leq s_1 < \cdots < s_J < k$ such that

$$I_j = \{s_j + 1, s_j + 2, \ldots, s_j + \ell_j\}\ (j = 1, \ldots, J) \tag{3.30}$$

$$\text{and } s_j + \ell_j \leq s_{j+1}\ (j = 1, \ldots, J-1).$$

We define the null hypothesis $H^o(I_1, \ldots, I_J)$ by

$$\begin{aligned} H^o(I_1, \ldots, I_J) : &\text{ for any } j \text{ such that } 1 \leq j \leq J \text{ and} \\ &\text{ for any } i, i' \in I_j,\ \mu_i = \mu_{i'} \text{ holds}. \end{aligned} \tag{3.31}$$

The elements of I_j are consecutive integers and $\ell_j = \#(I_j) \geq 2$. From (3.31), for any nonempty $V \subset \mathscr{U}$, there exist an integer J and some subsets $I_1, \ldots, I_J \subset \{1, \ldots, k\}$ satisfying (C4) such that

$$\bigwedge_{v \in V^+} H_v = H^o(I_1, \ldots, I_J). \tag{3.32}$$

Furthermore, $H^o(I_1, \ldots, I_J)$ is expressed as

$$H^o(I_1, \ldots, I_J) :\ \mu_{s_j+1} = \mu_{s_j+2} = \cdots = \mu_{s_j+\ell_j}\ (j = 1, \ldots, J). \tag{3.33}$$

Let us put

$$T^o(I_j) = \max_{s_j+1 \leq i < i' \leq s_j+\ell_j} T_{i'i} \quad (j = 1, \ldots, J),$$

where I_j is defined in (C4) and $T_{i'i}$ is defined by (3.15).

Corresponding to (3.16), (3.18), and (3.25), for ℓ such that $2 \leq \ell \leq k$, we put

$$D_1(t|\ell) := P\left(\max_{1 \leq i < i' \leq \ell} \frac{Z_{i'} - Z_i}{\sqrt{2}} \leq t \right), \tag{3.34}$$

$$TD_1(t|\ell, m) := P\left(\max_{1\le i<i'\le \ell} \frac{Z_{i'} - Z_i}{\sqrt{2U_E/m}} \le t\right)$$

and

$$td_1(\ell, m; \alpha) := \text{a solution of } t \text{ satisfying the equation } TD_1(t|\ell, m) = 1 - \alpha, \tag{3.35}$$

where Z_i and U_E are random variables used in (3.18).

Corresponding to (3.18), we have

$$TD_1(t|\ell, m) = \int_0^\infty D_1(ts|\ell)g(s|m)ds.$$

Then, we propose the stepwise procedure [3.3].

[3.3] Stepwise Procedure Based on One-Sided t-Test Statistics

For $H^o(I_1, \ldots, I_J)$ of (3.32), we set

$$M = M(I_1, \ldots, I_J) = \sum_{j=1}^{J} \ell_j. \tag{3.36}$$

For $\ell = \ell_1, \ldots, \ell_J$, we define $\alpha(M, \ell)$ by (1.19). By obeying the notation $td_1(\ell, m; \alpha)$, we get

$$TD_1(td_1(\ell, m; \alpha(M, \ell))|\ell, m) = 1 - \alpha(M, \ell) = (1 - \alpha)^{\ell/M}, \tag{3.37}$$

that is, $td_1(\ell, m; \alpha(M, \ell))$ is an upper $100\alpha(M, \ell)\%$ point of the distribution $TD_1(t|\ell, m)$.

(a) $J \ge 2$
Whenever $td_1\left(\ell_j, m; \alpha(M, \ell_j)\right) < T^o(I_j)$ holds for an integer j such that $1 \le j \le J$, we reject the hypothesis $\bigwedge_{v\in V^+} H_v$.

(b) $J = 1$ $(M = \ell_1)$
Whenever $td_1\,(M, m; \alpha) < T^o(I_1)$, we reject the hypothesis $\bigwedge_{v\in V^+} H_v$.

By using the methods of (a) and (b), when $\bigwedge_{v\in V^+} H_v$ is rejected for any V such that $(i, i') \in V \subset \mathscr{U}$, the null hypothesis $H_{(i,i')}$ is rejected as a multiple comparison test, where V^+ is defined by (3.28).

Theorem 3.1 *The test procedure [3.3] is a multiple comparison test of level α.*

Proof Refer to Shiraishi and Sugiura (2018). □

From (3.32), we find

$$\overline{\mathscr{H}^o} = \left\{ H^o(I_1, \ldots, I_J) \,\middle|\, \text{There exists } J \text{ such that } (C4),\right.$$

$$\bigcup_{j=1}^{J} I_j \subset \{1, \ldots, k\},\ \#(I_j) \geq 2\ (1 \leq j \leq J),$$

$$\left.\text{and } I_j \cap I_{j'} = \emptyset\ (1 \leq j < j' \leq J,\ J \geq 2) \text{ are satisfied.}\right\}$$

For $(i, i') \in \mathscr{U}$, let us put

$$\overline{\mathscr{H}^o}_{(i,i')} := \left\{ H^o(I_1, \ldots, I_J) \in \overline{\mathscr{H}^o} \mid \text{There exists } j \text{ such that } \{i, i'\} \subset I_j \text{ and } 1 \leq j \leq J \right\}.$$

Then we get

$$\overline{\mathscr{H}^o} = \bigcup_{(i,i') \in \mathscr{U}} \overline{\mathscr{H}^o}_{(i,i')} \quad \text{and} \quad H_{(i,i')},\ H_0 \in \overline{\mathscr{H}^o}_{(i,i')}.$$

Furthermore, we have, for $1 \leq i_1 \leq i_2 < i_2' \leq i_1' \leq k$,

$$\overline{\mathscr{H}^o}_{(i_1,i_1')} \subset \overline{\mathscr{H}^o}_{(i_2,i_2')}. \tag{3.38}$$

Therefore, by following (i) and (ii), we make a decision to reject or retain $H_{(i,i')}$ as a multiple comparison test of level α for $(i, i') \in \mathscr{U}$.

(i) Whenever all the elements of $\overline{\mathscr{H}^o}_{(i,i')}$ are rejected, $H_{(i,i')}$ is rejected.
(ii) Whenever there exists an element of $\overline{\mathscr{H}^o}_{(i,i')}$ that is not rejected, $H_{(i,i')}$ is not rejected.

For $k = 4$, all the elements $H^o(I_1, \ldots, I_J)$ of $\overline{\mathscr{H}^o}_{(1,2)}$ are stated in Table 3.1. From Table 3.1, in order to reject $H_{(1,2)}$ as a multiple comparison test, null hypotheses must be rejected. Whenever the following (3-1)–(3-4) are satisfied, the closed testing procedure of [3.3] rejects $H_{(1,2)}$ as a multiple comparison test of level α (Tables 3.2–3.6).

(3-1) $T^o(\{1, 2, 3, 4\}) = \max_{1 \leq i < i' \leq 4} T_{i'i} > td_1(4, m; \alpha)$.
(3-2) $T^o(\{1, 2\}) = T_{21} > td_1(2, m; \alpha(4, 2))$
or $T^o(\{3, 4\}) = T_{43} > td_1(2, m; \alpha(4, 2))$.
(3-3) $T^o(\{1, 2, 3\}) = \max_{1 \leq i < i' \leq 3} T_{i'i} > td_1(3, m; \alpha)$.
(3-4) $T^o(\{1, 2\}) = T_{21} > td_1(2, m; \alpha)$.

Table 3.1 When $k = 4$, in testing the null hypothesis $H_{(1,2)}$ as a multiple comparison, the null hypotheses $H(I_1, \ldots, I_J) \in \overline{\mathscr{H}^o}_{(1,2)}$ that are tested as a closed testing procedure

M	$H^o(I_1, \ldots, I_J)$
4	$H^o(\{1, 2, 3, 4\})$, $H^o(\{1, 2\}, \{3, 4\})$
3	$H^o(\{1, 2, 3\})$
2	$H^o(\{1, 2\})$

Table 3.2 When $k = 4$, the null hypotheses $H^o(I_1, \ldots, I_J) \in \overline{\mathscr{H}}^o_{(1,3)}$

M	$H^o(I_1, \ldots, I_J)$
4	$H^o(\{1, 2, 3, 4\})$
3	$H^o(\{1, 2, 3\})$

Table 3.3 When $k = 4$, the null hypotheses $H^o(I_1, \ldots, I_J) \in \overline{\mathscr{H}}^o_{(1,4)}$

M	$H^o(I_1, \ldots, I_J)$
4	$H^o(\{1, 2, 3, 4\})$

Table 3.4 When $k = 4$, the null hypotheses $H^o(I_1, \ldots, I_J) \in \overline{\mathscr{H}}^o_{(2,3)}$

M	$H^o(I_1, \ldots, I_J)$
4	$H^o(\{1, 2, 3, 4\})$
3	$H^o(\{1, 2, 3\})$, $H^o(\{2, 3, 4\})$
2	$H^o(\{2, 3\})$

Table 3.5 When $k = 4$, the null hypotheses $H^o(I_1, \ldots, I_J) \in \overline{\mathscr{H}}^o_{(2,4)}$

M	$H^o(I_1, \ldots, I_J)$
4	$H^o(\{1, 2, 3, 4\})$
3	$H^o(\{2, 3, 4\})$

Table 3.6 When $k = 4$, the null hypotheses $H^o(I_1, \ldots, I_J) \in \overline{\mathscr{H}}^o_{(3,4)}$

M	$H^o(I_1, \ldots, I_J)$
4	$H^o(\{1, 2, 3, 4\})$, $H^o(\{1, 2\}, \{3, 4\})$
3	$H^o(\{2, 3, 4\})$
2	$H^o(\{3, 4\})$

$$H^o(\{1, 2, 3, 4\}) : \ \mu_1 = \mu_2 = \mu_3 = \mu_4; \ J = 1, \ s_1 = 0, \ \ell_1 = 4$$
$$H^o(\{1, 2\}, \{3, 4\}) : \ \mu_1 = \mu_2, \ \mu_3 = \mu_4; \ J = 2, \ s_1 = 0, \ \ell_1 = 2, \ s_2 = 2, \ \ell_2 = 2$$
$$H^o(\{1, 2, 3\}) : \ \mu_1 = \mu_2 = \mu_3; \ J = 1, \ s_1 = 0, \ \ell_1 = 3$$
$$H^o(\{1, 2\}) = H_{(1,2)} : \ \mu_1 = \mu_2; \ J = 1, \ s_1 = 0, \ \ell_1 = 2$$

From a definition, we can verify that $td_1(\ell, m; \alpha) < td_1(k, m; \alpha)$ holds for ℓ such that $2 \leq \ell < k$. For $\alpha = 0.05, \ 0.01$, we give the values of $td_1(\ell; \alpha(M, \ell))$ in Tables 6.4 and 6.5 of Sect. 6.4, respectively. We limited attention to $m = 60$ and $2 \leq M \leq 10$. When $\ell = M = k$ is satisfied, $td_1(\ell, m; \alpha(M, \ell)) = td_1(k, m; \alpha)$ holds.

When $\alpha = 0.05, \ 0.01$, $m = 60$, $4 \leq k \leq 10$, from Tables 6.4 and 6.5, we find

$$td_1(\ell, m; \alpha(M, \ell)) < td_1(k, m; \alpha(k, k)) = td_1(k, m; \alpha) \tag{3.39}$$

for ℓ such that $2 \leq \ell < M \leq k$. By numerical calculation, we verify that (3.39) holds for $m = 50(2)150$, $\alpha = 0.05, \ 0.01$, and $3 \leq k \leq 10$. From the construction of the closed testing procedure [3.3] and the relation of (3.39), we get the following (i) and (ii). (i) The procedure [3.3] of level α rejects $H_{(i,i')}$ that is rejected by the simultaneous test [3.1] of level α. (ii) The simultaneous test [3.1] of level α does not always reject

$H_{(i,i')}$ that is rejected by the procedure [3.3] of level α. Hence, for $\alpha = 0.05,\ 0.01$, $3 \le k \le 10$ and $m = 50(2)150$, the closed testing procedure [3.3] is more powerful than the single-step simultaneous test [3.1].

We get Theorem 3.2 similar to Theorem 1.3.

Theorem 3.2 *Let $A^o_{(i,i')}$ be the event that $H_{(i,i')}$ is rejected by the procedure [3.3] as a multiple comparison of level α ($(i,i') \in \mathscr{U}$). Suppose that*

$$td_1\ (\ell, m; \alpha(M,\ell)) < td_1(k, m; \alpha) \tag{3.40}$$

is satisfied for any M such that $4 \le M \le k$ and any integer ℓ such that $2 \le \ell \le M-2$, where M is defined by (3.36). Then the following relations hold:

$$\bigcup_{(i,i')\in\mathscr{U}} A^o_{(i,i')} = \left\{ \max_{(i,i')\in\mathscr{U}} T_{i'i} > td_1(k,m;\alpha) \right\}, \tag{3.41}$$

$$A^o_{(i,i')} \supset \{T_{i'i} > td_1(k,m;\alpha)\} \quad \left((i,i') \in \mathscr{U}\right), \tag{3.42}$$

$$P\left(\bigcup_{(i,i')\in\mathscr{U}} A^o_{(i,i')} \right) = P\left(\max_{(i,i')\in\mathscr{U}} T_{i'i} > td_1(k,m;\alpha) \right), \tag{3.43}$$

$$P\left(A^o_{(i,i')}\right) \ge P\left(T_{i'i} > td_1(k,m;\alpha)\right) \quad \left((i,i') \in \mathscr{U}\right). \tag{3.44}$$

Proof Refer to Shiraishi and Sugiura (2015).

The left-hand side (l.h.s) of (3.43) is the probability that the procedure [3.3] rejects at least one of the hypotheses in $\mathscr{H}$. The right-hand side (r.h.s.) of (3.43) is the probability that the simultaneous test [3.1] rejects at least one of these hypotheses. The l.h.s. of (3.44) is the probability that the procedure [3.3] rejects $H_{(i,i')}$. The r.h.s. of (3.44) is the probability that the test [3.1] rejects $H_{(i,i')}$. The relation (3.44) means that the l.h.s. is greater than or equal to the r.h.s. For any $\boldsymbol{\mu}$, (3.43) and (3.44) hold.

We consider the $100(1-\alpha)\%$ confidence region for $\boldsymbol{\mu} = (\mu_1, \ldots, \mu_k)$ which is induced by the test procedure [3.3]. Suppose that the condition of Theorem 3.2 is satisfied. We set $\boldsymbol{X} = (X_{11}, \ldots, X_{1n_1}, \ldots, X_{k1}, \ldots, X_{kn_k})$. For any $(i,i') \in \mathscr{U}$, there exists $A^{o*}_{(i,i')}$ such that $A^o_{(i,i')} = \left\{\boldsymbol{X} \in A^{o*}_{(i,i')}\right\}$. From (3.41), we find

$$\bigcup_{(i,i')\in\mathscr{U}} \left\{\boldsymbol{X} \in A^{o*}_{(i,i')}\right\} = \left\{ \max_{(i,i')\in\mathscr{U}} T_{i'i} > td_1(k,m;\alpha) \right\}. \tag{3.45}$$

The direct product of $\boldsymbol{\mu}$ and $\mathbf{1}_n$ is denoted by $\boldsymbol{\mu} \otimes \mathbf{1}_n = (\mu_1 \mathbf{1}_{n_1}, \ldots, \mu_k \mathbf{1}_{n_k})$, where $\mathbf{1}_{n_i}$ is the row vector consisting of n_i ones. By replacing $\boldsymbol{X}$ with $\boldsymbol{X} - \boldsymbol{\mu} \otimes \mathbf{1}_n$ in (3.45), we have

$$\bigcup_{(i,i')\in\mathscr{U}} \left\{ \boldsymbol{X} - \boldsymbol{\mu} \otimes \mathbf{1}_n \in A^{o*}_{(i,i')} \right\} = \left\{ \max_{(i,i')\in\mathscr{U}} T_{i'i}(\boldsymbol{\mu}) > td_1(k, m; \alpha) \right\}. \tag{3.46}$$

From (3.46) and (3.43), we get

$$\begin{aligned} P_{\mu}\left(\bigcap_{(i,i')\in\mathscr{U}} \left\{ \boldsymbol{X} - \boldsymbol{\mu} \otimes \mathbf{1}_n \in \left(A^{o*}_{(i,i')}\right)^c \right\} \right) &= P_{\mu}\left(\max_{(i,i')\in\mathscr{U}} T_{i'i}(\boldsymbol{\mu}) \le td_1(k, m; \alpha) \right) \\ &= P_0\left(\max_{(i,i')\in\mathscr{U}} T_{i'i} \le td_1(k, m; \alpha) \right) \\ &= 1 - \alpha. \end{aligned} \tag{3.47}$$

Hence, (3.46) and (3.47) imply that the $100(1 - \alpha)\%$ confidence region for $\boldsymbol{\mu} = (\mu_1, \ldots, \mu_k)$ induced by the test procedure [3.3] becomes

$$\bigcap_{(i,i')\in\mathscr{U}} \left\{ \boldsymbol{\mu} \,\middle|\, \boldsymbol{X} - \boldsymbol{\mu} \otimes \mathbf{1}_n \in \left(A^{o*}_{(i,i')}\right)^c \right\} = \left\{ \boldsymbol{\mu} \,\middle|\, \max_{1\le i<i'\le k} T_{i'i}(\boldsymbol{\mu}) \le td_1(k, m; \alpha) \right\}. \tag{3.48}$$

Since the right-hand side of (3.48) is equal to (3.26), the $100(1 - \alpha)\%$ confidence region for $\boldsymbol{\mu} = (\mu_1, \ldots, \mu_k)$ induced by the test procedure [3.3] is equivalent to $100(1 - \alpha)\%$ simultaneous confidence intervals [3.2].

3.3 Closed Testing Procedures Under Unequal Sample Sizes

We do not suppose the condition (C3) of equal sample sizes . The discussions of (3.27)–(3.33) do not depend on the condition (C3).

For I_j of (3.30) and $j = 1, \ldots, J$, we define $\tilde{\mu}^*_{s_j+1}(I_j), \ldots, \tilde{\mu}^*_{s_j+\ell_j}(I_j)$ by $u_{s_j+1}, \ldots, u_{s_j+\ell_j}$ which minimize $\sum_{i\in I_j} \lambda_{ni}\left(u_i - \bar{X}_{i\cdot}\right)^2$ under simple order restrictions $u_{s_j+1} \le u_{s_j+2} \le \cdots \le u_{s_j+\ell_j}$, i.e.,

$$\sum_{i\in I_j} \lambda_{ni}\left(\tilde{\mu}^*_i(I_j) - \bar{X}_{i\cdot}\right)^2 = \min_{u_{s_j+1}\le\cdots\le u_{s_j+\ell_j}} \sum_{i\in I_j} \lambda_{ni}\left(u_i - \bar{X}_{i\cdot}\right)^2.$$

Corresponding to (3.3), we get

$$\tilde{\mu}^*_{s_j+r}(I_j) = \max_{s_j+1\le p\le s_j+r} \ \min_{s_j+r\le q\le s_j+\ell_j} \frac{\sum_{i=p}^{q} n_i \bar{X}_{i\cdot}}{\sum_{i=p}^{q} n_i} \quad (r = 1, \ldots, \ell_j).$$

We put

$$\bar{B}_1^2(I_j) := \frac{\sum_{i \in I_j} n_i \left(\tilde{\mu}_i^*(I_j) - \bar{X}_{\cdot\cdot}(I_j)\right)^2}{V_E}, \tag{3.49}$$

where

$$\bar{X}_{\cdot\cdot}(I_j) := \frac{\sum_{i \in I_j} \sum_{t=1}^{n_i} X_{it}}{N(I_j)}, \quad N(I_j) := \sum_{i \in I_j} n_i. \tag{3.50}$$

Let $P(L, \ell_j; \boldsymbol{\lambda}_n(I_j))$ be the probability that $\tilde{\mu}_{s_j+1}^*(I_j), \ldots, \tilde{\mu}_{s_j+\ell_j}^*(I_j)$ takes exactly L distinct values under H_0, where $\boldsymbol{\lambda}_n(I_j) := (n_{s_j+1}/n, n_{s_j+2}/n, \ldots, n_{s_j+\ell_j}/n)$. Then, from (3.12), for $t > 0$, under $H^o(I_1, \ldots, I_J)$ of (3.33), we get

$$\begin{aligned} P(\bar{B}_1^2(I_j) \geq t) &= P_0(\bar{B}_1^2(I_j) \geq t) \\ &= \sum_{L=2}^{\ell_j} P(L, \ell_j; \boldsymbol{\lambda}_n(I_j)) P\left((L-1)F_m^{L-1} \geq t\right). \end{aligned} \tag{3.51}$$

For a given α such that $0 < \alpha < 0.5$, we put

$$\bar{b}_1^2(\ell_j, \boldsymbol{\lambda}_n(I_j), m; \alpha) := \text{a solution of } t \text{ satisfying the equation } P_0(\bar{B}_1^2(I_j) \geq t) = \alpha. \tag{3.52}$$

We put

$$\bar{\chi}_1^2(I_j) := \frac{\sum_{i \in I_j} n_i \left(\tilde{\mu}_i^*(I_j) - \bar{X}_{\cdot\cdot}(I_j)\right)^2}{\sigma^2}.$$

We define $\breve{\mu}_1^*, \ldots, \breve{\mu}_{\ell_j}^*$ by

$$\sum_{i=1}^{\ell_j} \lambda_{s_j+i} \left(\breve{\mu}_i^* - Z_i\right)^2 = \min_{u_1 \leq \cdots \leq u_{\ell_j}} \sum_{i=1}^{\ell_j} \lambda_{s_j+i} (u_i - Z_i)^2$$

where $Z_1, \ldots, Z_{\ell_j}$ are independent and $Z_i \sim N(0, 1/\lambda_{s_j+i})$ $(i = 1, \ldots, \ell_j)$. Let $P(L, \ell_j; \boldsymbol{\lambda}(I_j))$ be the probability that $\breve{\mu}_1^*, \ldots, \breve{\mu}_{\ell_j}^*$ takes exactly L distinct values, where $\boldsymbol{\lambda}(I_j) := (\lambda_{s_j+1}, \ldots, \lambda_{s_j+\ell_j})$. Then, for $t > 0$, under the condition (C1), we get

$$\lim_{n\to\infty} P_0\left(\bar{\chi}_1^2(I_j) \geq t\right) = \sum_{L=2}^{\ell_j} P(L, \ell_j; \boldsymbol{\lambda}(I_j)) P\left(\chi_{L-1}^2 \geq t\right). \tag{3.53}$$

Furthermore, for $t > 0$, under the condition (C1),

$$\lim_{n\to\infty} P_0(\bar{B}_1^2(I_j) \geq t) = \lim_{n\to\infty} P_0\left(\bar{\chi}_1^2(I_j) \geq t\right) \tag{3.54}$$

holds. For a given α such that $0 < \alpha < 0.5$, we put

$$\bar{c}_1^2\left(\ell_j, \boldsymbol{\lambda}(I_j); \alpha\right) := \text{a solution of } t \text{ satisfying the equation } \lim_{n\to\infty} P_0\left(\bar{\chi}_1^2(I_j) \geq t\right) = \alpha. \tag{3.55}$$

Under (C1), we have

$$\lim_{n\to\infty} \bar{b}_1^2(\ell_j, \boldsymbol{\lambda}_n(I_j), m; \alpha) = \bar{c}_1^2\left(\ell_j, \boldsymbol{\lambda}(I_j); \alpha\right).$$

Then, we propose the stepwise procedure [3.4].

[3.4] Stepwise procedure based on $\bar{B}_1^2$ statistics

For $H^o(I_1, \ldots, I_J)$ of (3.32) and for $\ell = \ell_1, \ldots, \ell_J$, we define M and $\alpha(M, \ell)$ by (3.36) and (1.19), respectively.

(a) $J \geq 2$
Whenever $\bar{b}_1^2\left(\ell_j, \boldsymbol{\lambda}_n(I_j), m; \alpha(M, \ell_j)\right) < \bar{B}_1^2(I_j)$ holds for an integer j such that $1 \leq j \leq J$, we reject the hypothesis $\bigwedge_{v\in V^+} H_v$.

(b) $J = 1$ $(M = \ell_1)$
Whenever $\bar{b}_1^2\left(\ell_1, \boldsymbol{\lambda}_n(I_1), m; \alpha\right) < \bar{B}_1^2(I_1)$, we reject the hypothesis $\bigwedge_{v\in V^+} H_v$.

By using the methods of (a) and (b), when $\bigwedge_{v\in V^+} H_v$ is rejected for any V such that $(i, i') \in V \subset \mathcal{U}$, the null hypothesis $H_{(i,i')}$ is rejected as a multiple comparison test, where V^+ is defined by (3.28).

Theorem 3.3 *The test procedure [3.4] is a multiple comparison test of level α.*

Proof The proof is similar to that of Theorem 5.5 of Shiraishi and Sugiura (2018). □

References

Barlow RE, Bartholomew DJ, Bremner JM, Brunk HD (1972) Statistical Inference under Order Restrictions. Wiley, London

Hayter AJ (1990) A one-sided Studentized range test for testing against a simple ordered alternative. J. Amer. Statist. Assoc. 85:778–785

Miwa T, Hayter AJ, Liu W (2000) Calculations of level probabilities for normal random variables with unequal variances with applications to Bartholomew's test in unbalanced one-way models. Computational Statistics and Data Analysis 34:17–32

Robertson T, Wright FT, Dykstra RL (1988) Order Restricted Statistical Inference. Wiley, New York

Shiraishi T (2014) Closed Testing Procedures in Multi-Sample Models under a Simple Order Restriction. J. Japan Statistical Society. Japanese Issue 43:215–245 (in Japanese)

Shiraishi T, Sugiura H (2015) The Upper $100\alpha^\star$th Percentiles of the Distributions Used in Multiple Comparison Procedures Under a Simple Order Restriction. J. Japan Statistical Society. Japanese Issue 44:271–314 (in Japanese)

Shiraishi T, Sugiura H (2018) Theory of Multiple Comparison Procedures and Its Computation. Kyoritsu-Shuppan Co., Ltd (in Japanese)

Chapter 4
Nonparametric Procedures Based on Rank Statistics

Abstract We consider distribution-free multiple comparison procedures among mean effects in homoscedastic k-sample models. We propose closed testing procedures based on the maximum values of some two-sample Wilcoxon test statistics and based on Kruskal–Wallis test statistics. The results reveal that the proposed procedures are more powerful than single-step procedures and the REGW-type tests. Furthermore, under simple order restrictions of k means, we discuss distribution-free multiple comparison procedures.

4.1 Introduction

We consider homoscedastic k-sample models. $(X_{i1}, \ldots, X_{in_i})$ is a random sample of size n_i from the ith population with mean μ_i $(i = 1, \ldots, k)$, that is, $P(X_{ij} \leq x) = F(x - \mu_i)$, where $F(x)$ is an absolutely continuous distribution function. Furthermore, X_{ij}'s are assumed to be independent. $\int_{-\infty}^{\infty} x dF(x) = 0$ holds. Let R_{ij} be the rank of X_{ij} among all n observations $\{X_{ij} \mid 1 \leq j \leq n_i,\ 1 \leq i \leq k\}$, where n is defined in (1.2). To test the null hypothesis H_0 of homogeneity of k means given by (1.3), Kruskal and Wallis (1952) proposed a distribution-free procedure based on

$$T_R := \frac{12}{n(n+1)} \sum_{i=1}^{k} n_i \left(\bar{R}_{i\cdot} - \frac{n+1}{2} \right)^2,$$

where $\bar{R}_{i\cdot} := \frac{1}{n_i} \sum_{j=1}^{n_i} R_{ij}$. If the condition (C1) of (2.5) is satisfied, under H_0, we find, as n tends to infinity,

$$Z_R \xrightarrow{\mathscr{L}} \chi^2_{k-1}, \tag{4.1}$$

where $\xrightarrow{\mathscr{L}}$ denotes convergence in law and χ^2_{k-1} denotes the χ^2 random variable with $k-1$ degrees of freedom. We reject H_0 at level α if $Z_R > \chi^2_{k-1}(\alpha)$, where $\chi^2_{k-1}(\alpha)$ denotes the upper $100\alpha\%$ point of χ^2-distribution with degrees of freedom $k-1$.

T.-a. Shiraishi et al., *Pairwise Multiple Comparisons*,
JSS Research Series in Statistics,
https://doi.org/10.1007/978-981-15-0066-4_4

We consider test procedures for all-pairwise comparisons of

$$\left\{\text{the null hypothesis } H_{(i,i')} : \mu_i = \mu_{i'} \text{ vs. the alternative } H^A_{(i,i')} : \mu_i \neq \mu_{i'} \,\middle|\, (i,i') \in \mathscr{U}\right\},$$

where $\mathscr{U}$ is defined by (1.5).

Steel (1960) and Dwass (1960) proposed single-step procedures as multiple comparison tests of level α. Shiraishi (2011a) proposed distribution-free closed testing procedures. Our results show that (i) the proposed multistep procedures are more powerful than the single-step procedures of Steel (1960) and Dwass (1960) and the REGW (Ryan–Einot–Gabriel–Welsh) type rank tests, and (ii) confidence regions induced by the multistep procedures are equivalent to distribution-free simultaneous confidence intervals.

4.2 The Single-Step Procedures

We put $N_{i'i} := n_i + n_{i'}$. Let $R^{(i',i)}_{i'\ell}$ denote the rank of $X_{i'\ell}$ among $N_{i'i}$ observations $X_{i1}, \ldots, X_{in_i}, X_{i'1}, \ldots, X_{i'n_{i'}}$ $(i < i')$, and let us put

$$\widehat{T}_{i'i} := \sum_{\ell=1}^{n_{i'}} R^{(i',i)}_{i'\ell} - \frac{n_{i'}(N_{i'i}+1)}{2}.$$

Then the mean and variance of $\widehat{T}_{i'i}$ under H_0 of (1.3) are, respectively, given by

$$E_0(\widehat{T}_{i'i}) = 0 \text{ and } V_0(\widehat{T}_{i'i}) = \frac{n_i n_{i'}(N_{i'i}+1)}{12}.$$

Hence, we put

$$\widehat{Z}_{i'i} := \frac{\widehat{T}_{i'i}}{\sigma_{i'in}}, \tag{4.2}$$

where

$$\sigma_{i'in} := \sqrt{\frac{n_i n_{i'}(N_{i'i}+1)}{12}}. \tag{4.3}$$

For given α such that $0 < \alpha < 1$, we put

$$ae(k, n_1, \ldots, n_k; \alpha) := \text{a solution of } t \text{ satisfying the two inequalities} \tag{4.4}$$

$$P_0\left(\max_{1\le i<i'\le k} |\widehat{Z}_{i'i}| > t\right) \le \alpha \text{ and } P_0\left(\max_{1\le i<i'\le k} |\widehat{Z}_{i'i}| \ge t\right) > \alpha.$$

We have the distribution-free single-step tests based on rank statistics proposed by Shiraishi (2011a).

[4.1] Distribution-Free Single-Step Tests Based on Rank Statistics

The distribution-free simultaneous test of level α for all-pairwise comparisons of $\left\{\text{the null hypothesis } H_{(i,i')} \text{ vs. the alternative } H^A_{(i,i')} \,\middle|\, (i,i') \in \mathscr{U}\right\}$ consists in rejecting $H_{(i,i')}$ for $(i,i') \in \mathscr{U}$ such that $|\widehat{Z}_{i'i}| > ae(k, n_1, \ldots, n_k; \alpha)$

We introduce the distribution function of $A^*(t)$.

$$A^*(t) := \sum_{j=1}^{k} \int_{-\infty}^{\infty} \prod_{\substack{i=1 \\ i \neq j}}^{k} \left\{ \Phi\left(\sqrt{\frac{\lambda_i}{\lambda_j}} \cdot x \right) - \Phi\left(\sqrt{\frac{\lambda_i}{\lambda_j}} \cdot x - \sqrt{\frac{\lambda_i + \lambda_j}{\lambda_j}} \cdot t \right) \right\} d\Phi(x).$$

Then we get Theorem 4.1.

Theorem 4.1 *Suppose that (C1) of (2.5) is satisfied. Then, for $t > 0$,*

$$A(t) \le \lim_{n \to \infty} P_0 \left(\max_{(i,i') \in \mathscr{U}} |\widehat{Z}_{i'i}| \le t \right) \le A^*(t) \tag{4.5}$$

holds, where $A(t)$ is defined by (2.8), $P_0(\cdot)$ stands for probability measure under the null hypothesis H_0. When $\lambda_1 = \cdots = \lambda_k = 1/k$ is satisfied, both of the inequalities of (4.5) become an equality.

Proof Refer to Shiraishi (2011b). □

We have the single-step tests proposed by Steel (1960) and Dwass (1960).

[4.2] Asymptotic Single-Step Tests Based on Rank Statistics

The Steel–Dwass simultaneous test of level α for all-pairwise comparisons of $\left\{\text{the null hypothesis } H_{(i,i')} \text{ vs. the alternative } H^A_{(i,i')} \,\middle|\, (i,i') \in \mathscr{U}\right\}$ consists in rejecting $H_{(i,i')}$ for $(i,i') \in \mathscr{U}$ such that $|\widehat{Z}_{i'i}| > a(k;\alpha)$, where $a(k;\alpha)$ is defined by (2.10). From the left inequality of (4.5), we find that the Steel–Dwass simultaneous test is conservative. Under the condition of $\max_{1 \le i \le k} n_i / \min_{1 \le i \le k} n_i \le 2$, Shiraishi (2007) found that the values of $A^*(t) - A(t)$ are nearly equal to 0 for various values of t from numerical integration. Therefore, the conservativeness of the Steel–Dwass method is small.

[4.3] Distribution-Free Simultaneous Confidence Intervals

We put $\theta_{i'i} \equiv \mu_{i'} - \mu_i$. Let $R^{(i',i)}_{i\ell}(\theta_{i'i})$ denote the rank of $X_{i'\ell} - \theta_{i'i}$ among $N_{i'i}$ observations $X_{i1}, \ldots, X_{in_i}, X_{i'1} - \theta_{i'i}, \ldots, X_{i'n_{i'}} - \theta_{i'i}$ $(i < i')$, and let us put

$$\widehat{T}_{i'i}(\theta_{i'i}) := \sum_{\ell=1}^{n_{i'}} R^{(i',i)}_{i'\ell}(\theta_{i'i}) - \frac{n_{i'}(N_{i'i}+1)}{2}.$$

Furthermore, we put

$$\widehat{Z}_{i'i}(\theta_{i'i}) := \frac{\widehat{T}_{i'i}(\theta_{i'i})}{\sigma_{i'in}}. \tag{4.6}$$

Let

$$\mathscr{D}_{(1)}^{(i',i)} \le \mathscr{D}_{(2)}^{(i',i)} \le \cdots \le \mathscr{D}_{(n_{i'}n_i)}^{(i',i)}$$

be ordered statistics of $n_i n_{i'}$ observations $\{X_{i'\ell'} - X_{i\ell} \mid \ell' = 1, \dots, n_{i'}, \ell = 1, \dots n_i\}$. Then from Shiraishi (2011b), we have the relations

$$\begin{aligned} 1 - \alpha &\le P_0\left(\max_{1 \le i < i' \le k} |\widehat{Z}_{i'i}| \le ae(k, n_1, \dots, n_k; \alpha) \right) \\ &= P\left(\max_{1 \le i < i' \le k} |\widehat{Z}_{ii'}(\theta_{i'i})| \le ae(k, n_1, \dots, n_k; \alpha) \right) \\ &= P\left(\mathscr{D}_{(\lceil ae_{i'i} \rceil)}^{(i',i)} \le \theta_{i'i} < \mathscr{D}_{(\lfloor n_{i'}n_i - ae_{i'i} \rfloor + 1)}^{(i',i)},\ 1 \le i < i' \le k \right), \end{aligned}$$

where

$$ae_{i'i} := -\sigma_{i'in} \cdot ae(k, n_1, \dots, n_k; \alpha) + \frac{n_{i'}n_i}{2},$$

the symbol $\lceil x \rceil$ denotes the smallest integer that is greater than or equal to x, and the symbol $\lfloor x \rfloor$ denotes the largest integer not exceeding x.

Hence, distribution-free $100(1 - \alpha)\%$ simultaneous confidence intervals for all-pairwise $\{\mu_{i'} - \mu_i \mid (i, i') \in \mathscr{U}\}$ are given by the following expression:

$$\mathscr{D}_{(\lceil ae_{i'i} \rceil)}^{(i',i)} \le \mu_{i'} - \mu_i < \mathscr{D}_{(\lfloor n_{i'}n_i - ae_{i'i} \rfloor + 1)}^{(i',i)} \qquad ((i, i') \in \mathscr{U}). \tag{4.7}$$

Next we give the asymptotic simultaneous confidence intervals.

[4.4] Asymptotic Simultaneous Confidence Intervals

Under the condition (C1), for $t \ge 0$,

$$\lim_{n \to \infty} P\left(\max_{(i,i') \in \mathscr{U}} |\widehat{Z}_{i'i}(\theta_{i'i})| \le t \right) = \lim_{n \to \infty} P_0\left(\max_{(i,i') \in \mathscr{U}} |\widehat{Z}_{i'i}| \le t \right) \tag{4.8}$$

holds. We get Theorem 4.2.

Theorem 4.2 *Suppose that the condition (C1) is satisfied. Then asymptotic* $100(1 - \alpha)\%$ *simultaneous confidence intervals for all-pairwise* $\{\mu_{i'} - \mu_i \mid (i, i') \in \mathscr{U}\}$ *are given by the following expression:*

$$\mathscr{D}_{(\lceil a_{i'i} \rceil)}^{(i',i)} \le \mu_{i'} - \mu_i < \mathscr{D}_{(\lfloor n_{i'}n_i - a_{i'i} \rfloor + 1)}^{(i',i)} \qquad ((i, i') \in \mathscr{U}), \tag{4.9}$$

where

$$a_{i'i} := \frac{n_{i'}n_i}{2} - \sigma_{i'in} \cdot a(k; \alpha).$$

Proof Refer to Shiraishi (2011b). □

Critchlow and Fligner (1991) used the other expressions which are equivalent to (4.7) and (4.9).

4.3 Closed Testing Procedures

We can discuss (1.14)–(1.18) for the nonparametric models. For disjoint sets $I_1, \ldots, I_J$ defined in (1.17), we put

$$\widehat{Z}(I_j) := \max_{i<i',\ i,i' \in I_j} |\widehat{Z}_{i'i}| \ (j = 1, \ldots, J).$$

Then, we propose the stepwise procedure [4.5].

[4.5] Stepwise Procedure Based on Rank Test Statistics

For M and $\ell = \ell_1, \ldots, \ell_J$ defined in (1.18), we define $\alpha(M, \ell)$ by (1.19).

(a) $J \geq 2$
Whenever there exists an integer j such that $1 \leq j \leq J$ and $a\left(\ell_j; \alpha(M, \ell_j)\right) \leq \widehat{Z}(I_j)$ are satisfied, we reject the hypothesis $\bigwedge_{\mathbf{v} \in V} H_{\mathbf{v}}$, where $a(\ell; \alpha(M, \ell))$ is an upper $100\alpha(M, \ell)\%$ point of the distribution $A(t|\ell)$ defined by (2.10).

(b) $J = 1$ $(M = \ell_1)$
Whenever $a\,(M; \alpha) \leq \widehat{Z}(I_1)$, we reject the hypothesis $\bigwedge_{\mathbf{v} \in V} H_{\mathbf{v}}$.

By using the methods of (a) and (b), when $\bigwedge_{\mathbf{v} \in V} H_{\mathbf{v}}$ is rejected for any V such that $(i, i') \in V \subset \mathscr{U}$, the null hypothesis $H_{(i,i')}$ is rejected as a multiple comparison test.

Theorem 4.3 *The test procedure [4.5] is asymptotically a multiple comparison test of level* α.

Proof Refer to Shiraishi (2011a). □

We get Theorem 4.4 similar to Theorem 1.3.

Theorem 4.4 *Let* $\hat{A}_{(i,i')}$ *be the event that* $H_{(i,i')}$ *is rejected by the procedure [4.5] as a multiple comparison of level* α $((i, i') \in \mathscr{U})$. *Suppose that*

$$a\,(\ell; \alpha(M, \ell)) < a(k; \alpha) \tag{4.10}$$

is satisfied for any M *such that* $4 \leq M \leq k$ *and any integer* ℓ *such that* $2 \leq \ell \leq M - 2$. *Then the following relations hold:*

$$P\left(\bigcup_{(i,i')\in\mathscr{U}} \hat{A}_{(i,i')}\right) = P\left(\max_{(i,i')\in\mathscr{U}} |\hat{Z}_{i'i}| > a(k;\alpha)\right),$$

$$P\left(\hat{A}_{(i,i')}\right) \geq P\left(|\hat{Z}_{i'i}| > a(k;\alpha)\right) \quad ((i,i') \in \mathscr{U}).$$

From Tables 2.2 and 2.3, we find that (4.10) is satisfied for $\alpha = 0.05,\ 0.01$.

4.4 Multiple Comparisons Under Simple Order Restrictions

In the model of Sect. 4.1, we add the condition of simple order restrictions (3.1). We consider test procedures for all-pairwise comparisons of

$$\left\{\text{the null hypothesis } H_{(i,i')} : \mu_i = \mu_{i'} \text{ vs. the alternative } H^{OA}_{(i,i')} : \mu_i < \mu_{i'} \,\middle|\, (i,i') \in \mathscr{U}\right\}.$$

4.4.1 Multiple Comparisons Under Equal Sample Sizes

We add the condition (C3) of equal sample sizes throughout this subsection appeared in (3.14). Then $\widehat{Z}_{i'i}$ of (4.2) becomes

$$\widehat{Z}_{i'i} = \frac{\sum_{\ell=1}^{n_1} R_{i'\ell}^{(i,i')} - \frac{n_1(2n_1+1)}{2}}{\sqrt{\frac{n_1^2(2n_1+1)}{12}}}. \tag{4.11}$$

Then we get Theorem 4.4.

Theorem 4.5 *For $t > 0$,*

$$\lim_{n\to\infty} P_0\left(\max_{(i,i')\in\mathscr{U}} \widehat{Z}_{i'i} \leq t\right) = D_1(t) \tag{4.12}$$

holds, where $D_1(t)$ is given by (3.16).

Proof Refer to Theorem 5.3 of Shiraishi and Sugiura (2018). □

For given α such that $0 < \alpha < 1$, we put

$$d_1(k;\alpha) := \text{a solution of } t \text{ satisfying the equation } D_1(t) = 1-\alpha. \tag{4.13}$$

Then we have the single-step tests [4.6] and the simultaneous confidence intervals [4.7] proposed by Shiraishi (2014).

[4.6] Single-Step Tests Based on Rank Statistics

The simultaneous test of level α for all-pairwise comparisons of $\left\{ \text{the null hypothesis } H_{(i,i')} \text{ vs. the alternative } H^{OA}_{(i,i')} \,\middle|\, (i,i') \in \mathscr{U} \right\}$, consists in rejecting $H_{(i,i')}$ for $(i,i') \in \mathscr{U}$ such that $\widehat{Z}_{i'i} > d_1(k;\alpha)$.

[4.7] Simultaneous Confidence Intervals

Let

$$\mathscr{D}^{(i',i)}_{(1)} \le \mathscr{D}^{(i',i)}_{(2)} \le \cdots \le \mathscr{D}^{(i',i)}_{(n_1^2)}$$

be ordered statistics of n_1^2 observations $\{X_{i'\ell'} - X_{i\ell} \mid \ell' = 1,\ldots,n_1, \ell = 1,\ldots n_1\}$. Then asymptotic $100(1-\alpha)\%$ simultaneous confidence intervals for all-pairwise $\{\mu_{i'} - \mu_i | \, (i,i') \in \mathscr{U}\}$ are given by

$$\mathscr{D}^{(i',i)}_{(\lceil a^*_{i'i} \rceil)} \le \mu_{i'} - \mu_i < +\infty \quad ((i,i') \in \mathscr{U}), \tag{4.14}$$

where

$$a^*_{i'i} := -\sqrt{\frac{n_1^2(2n_1+1)}{12}}d_1(k;\alpha) + \frac{n_1^2}{2}.$$

We can discuss (3.27)–(3.33) for the nonparametric models. For I_j of (3.30) and $j = 1,\ldots,J$, we put

$$\widehat{Z}^o(I_j) := \max_{s_j+1 \le i < i' \le s_j+\ell_j} \widehat{Z}_{i'i}.$$

[4.8] Stepwise Procedure Based on One-Sided Wilcoxon Test Statistics

For given α such that $0 < \alpha < 1$, we put

$$d_1(\ell;\alpha) := \text{a solution of } t \text{ satisfying the equation } D_1(t|\ell) = 1-\alpha, \tag{4.15}$$

where $D_1(t|\ell)$ is defined by (3.34). For $H^o(I_1,\ldots,I_J)$ of (3.32), we define M by (3.36). For $\ell = \ell_1,\ldots,\ell_J$, we define $\alpha(M,\ell)$ by (1.19). By obeying the notation $d_1(\ell;\alpha)$, we get

$$D_1(d_1(\ell;\alpha(M,\ell))|\ell) = 1-\alpha(M,\ell) = (1-\alpha)^{\ell/M}, \tag{4.16}$$

that is, $d_1(\ell;\alpha(M,\ell))$ is an upper $100\alpha(M,\ell)\%$ point of the distribution $D_1(t|\ell)$.

(a) $J \ge 2$
Whenever $d_1\left(\ell_j;\alpha(M,\ell_j)\right) < \widehat{Z}^o(I_j)$ holds for an integer j such that $1 \le j \le J$, we reject the hypothesis $\bigwedge_{\mathbf{v}\in V^+} H_{\mathbf{v}}$.

(b) $J = 1$ $(M = \ell_1)$
Whenever $d_1(M; \alpha) < \widehat{Z}^o(I_1)$, we reject the hypothesis $\bigwedge_{\mathbf{v} \in V^+} H_{\mathbf{v}}$.

By using the methods of (a) and (b), when $\bigwedge_{\mathbf{v} \in V^+} H_{\mathbf{v}}$ is rejected for any V such that $(i, i') \in V \subset \mathscr{U}$, the null hypothesis $H_{(i,i')}$ is rejected as a multiple comparison test, where V^+ is defined by (3.28).

Theorem 4.6 *Under the condition (C1), the test procedure [4.8] is an asymptotic multiple comparison test of level* α.

Proof Refer to Shiraishi and Sugiura (2018). □

4.4.2 Multiple Comparisons Under Unequal Sample Sizes

We do not suppose the condition (C3) of equal sample sizes. The discussions of (3.27)–(3.33) do not depend on the condition (C3). For I_j of (3.30), we define $R_{i\ell}(I_j)$ by the rank of $X_{i\ell}$ among $N(I_j)$ observations $\{X_{i\ell} \mid \ell = 1, \ldots, n_i, \ i \in I_j\}$, and we put

$$\bar{R}_{i\cdot}(I_j) = \frac{1}{n_i} \sum_{\ell=1}^{n_i} R_{i\ell}(I_j) \quad (i \in I_j),$$

where $N(I_j)$ is defined in (3.50). Furthermore, let us put

$$\bar{R}^*_{s_j+r\cdot}(I_j) = \max_{s_j+1 \le p \le s_j+r} \ \min_{s_j+r \le q \le s_j+\ell_j} \frac{\sum_{m=p}^{q} n_m \bar{R}_{m\cdot}(I_j)}{\sum_{m=p}^{q} n_m} \quad (r = 1, \ldots, \ell_j).$$

Then from (3.3),

$$\sum_{i \in I_j} \lambda_{ni} (\bar{R}^*_{i\cdot}(I_j) - \bar{R}_{i\cdot}(I_j))^2 = \min_{u_{s_j+1} \le \cdots \le u_{s_j+\ell_j}} \sum_{i \in I_j} \lambda_{ni} (u_i - \bar{R}_{i\cdot}(I_j))^2$$

holds. We put

$$\widehat{Z}_1^2(I_j) := \frac{12}{N(I_j)\{N(I_j)+1\}} \sum_{i \in I_j} n_i \left(\bar{R}^*_{i\cdot}(I_j) - \frac{N(I_j)+1}{2} \right)^2.$$

By using $\widehat{Z}_1^2(I_j)$ $(j = 1, \ldots, J)$, we propose the distribution-free closed testing procedure [4.9].

[4.9] Stepwise Procedure Based on $\bar{\chi}^2$-Rank Test Statistics

For $H^o(I_1, \ldots, I_J)$ of (3.32), we define M by (3.36). For $\ell = \ell_1, \ldots, \ell_J$, we define $\alpha(M, \ell)$ by (1.19).

(a) $J \geq 2$
Whenever $\bar{c}_1^2\left(\ell_j, \boldsymbol{\lambda}(I_j); \alpha(M, \ell_j)\right) < \widehat{Z}_1^2(I_j)$ holds for an integer j such that $1 \leq j \leq J$, we reject the hypothesis $\bigwedge_{\mathbf{v} \in V^+} H_{\mathbf{v}}$, where $\bar{c}_1^2\left(\ell_j, \boldsymbol{\lambda}(I_j); \alpha\right)$ is defined by (3.55).
(b) $J = 1 \ (M = \ell_1)$
Whenever $\bar{c}_1^2\left(\ell_1, \boldsymbol{\lambda}(I_1); \alpha\right) < \widehat{Z}_1^2(I_1)$, we reject the hypothesis $\bigwedge_{\mathbf{v} \in V^+} H_{\mathbf{v}}$.

By using the methods of (a) and (b), when $\bigwedge_{\mathbf{v} \in V^+} H_{\mathbf{v}}$ is rejected for any V such that $(i, i') \in V \subset \mathscr{U}$, the null hypothesis $H_{(i,i')}$ is rejected as a multiple comparison test.

Theorem 4.7 *Under the condition (C1), the test procedure [4.9] is an asymptotic multiple comparison test of level α.*

Proof Refer to Shiraishi and Sugiura (2018). □

References

Critchlow DE, Fligner MA (1991) Nonparametric multiple comparisons in the one-way analysis of variance. Commun Stat Ser A 20:127–139

Dwass M (1960) Some k-sample rank order tests, contributions to probability and statistics. Stanford University Press, Stanford, pp 198–202

Kruskal WH, Wallis WA (1952) Use of ranks in one-criterion variance analysis. J Am Stat Assoc 47:583–621

Shiraishi T (2007) The upper bound for the distribution of Tukey-Kramer's statistic. Bull Comput Stat Jpn 19:77–87 (in Japanese)

Shiraishi T (2011a) Closed testing procedures for pairwise comparisons in multi-sample models. Biom Soc Jpn 32:33–47 (in Japanese)

Shiraishi T (2011b) Multiple comparison procedures under continuous distributions. Kyoritsu-Shuppan Co., Ltd. (in Japanese)

Shiraishi T (2014) Closed testing procedures in multi-sample models under a simple order restriction. J Jpn Stat Soc Jpn Issue 43:215–245 (in Japanese)

Shiraishi T, Sugiura H (2018) Theory of multiple comparison procedures and its computation. Kyoritsu-Shuppan Co., Ltd. (in Japanese)

Steel RGD (1960) A rank sum test for comparing all pairs of treatments. Technometrics 2:197–207

Chapter 5
Comparison of Simulated Power Among Multiple Comparison Tests

Abstract In this chapter, we perform a Monte Carlo simulation to investigate the performance of several procedures discussed in Chaps. 1–4. Proposed procedures [1.5] and [3.4] demonstrate the best performance in each situation.

5.1 Introduction

We investigate the performance of several procedures presented in Chaps. 1–4. We then present the results of the simulations to evaluate the performance of the procedures, each of which is based on 100,000 Monte Carlo replicates.

We examine two types of procedures: all-pairwise comparisons with no restriction (see Chap. 1) and all-pairwise comparisons with simple order restrictions (see Chap. 3). The former type has (i) single-step tests based on t-statistics (see [1.1]), that is, the Tukey–Kramer method, denoted as [T1], (ii) the REGW method, which is denoted as [R2] (see [1.4]), (iii) stepwise procedure based on t-statistics, as [T3] (see [1.3]), and (iv) stepwise procedure based on F-statistics, as [F4] (see [1.5]).

The latter type comprises (v) single-step tests based on one-sided t-test statistics (see [3.1]), that is, the Hayter method, denoted as [H5], (vi) a stepwise procedure based on one-sided t-test statistics, as [T6] (see [3.3]), and (vii) a stepwise procedure based on $\bar{B}^2$-statistics, as [B7] (see [3.4]).

5.2 Simulation Settings

We simulate procedures under the null hypothesis H_0 and alternative hypotheses.

Alternative hypotheses comprise two types.

For $k = 4$, alternative hypotheses are the following ones:

$$H^{A1} : \mu_i = i\Delta/5 \quad (i = 1, 2, 3, 4),$$

$$H^{A2} : \mu_1 = \mu_2 = 0, \mu_3 = \mu_4 = \Delta/5.$$

T.-a. Shiraishi et al., *Pairwise Multiple Comparisons*,
JSS Research Series in Statistics,
https://doi.org/10.1007/978-981-15-0066-4_5

For $k = 5$, alternative hypotheses are the following ones:

$$H^{A1} : \mu_i = i\Delta/5 \quad (i = 1, 2, 3, 4, 5),$$

$$H^{A2} : \mu_1 = \mu_2 = \mu_3 = 0, \mu_4 = \mu_5 = \Delta/5.$$

5.3 Simulation Results

At first, we confirm familywise error rate for those procedures for the null hypothesis H_0.

Tables 5.1 and 5.2 show the results of familywise error rate for all-pairwise comparisons for $n = 16$, $\mu_i = 0$, and $k = 3, 4, 5$.

In this situation, we can use Tables 1.3 and 1.4 in simulations. The other necessary tables are described in Chap. 6.

We find that all procedures control the familywise error at level α in Tables 5.1 and 5.2.

In Tables 5.1 and 5.2, values of [T1], [R2], and [T3] are equal because the first steps of these procedures are the same. In the same manner, the values of [H5] and [T6] are equal.

At second, we investigate the all-pairs power, which is shown in Ramsey (1978), on alternative hypotheses. All-pairs power is defined by the probability that all hypotheses which are false are rejected.

Tables 5.3, 5.4, 5.5, and 5.6 show the results of all-pairs power for all-pairwise comparisons for $(k, n) = (4, 16), (5, 13)$. We use $\Delta = 3, 4, 5$ for $k = 4$ and $\Delta = 4, 5, 6$ for $k = 5$, respectively.

Table 5.1 Values of familywise error rate for [T1]–[B7] with $n = 16$, $\alpha = 0.05$, and H_0

$k\backslash$ procedure	[T1]	[R2]	[T3]	[F4]	[H5]	[T6]	[B7]
3	0.0499	0.0499	0.0499	0.0495	0.0496	0.0496	0.0493
4	0.0495	0.0495	0.0495	0.0482	0.0500	0.0500	0.0496
5	0.0499	0.0499	0.0499	0.0432	0.0497	0.0497	0.0503

Table 5.2 Values of familywise error rate for [T1]–[B7] with $n = 16$, $\alpha = 0.01$, and H_0

$k\backslash$ procedure	[T1]	[R2]	[T3]	[F4]	[H5]	[T6]	[B7]
3	0.0099	0.0099	0.0099	0.0099	0.0098	0.0098	0.0096
4	0.0097	0.0097	0.0097	0.0092	0.0093	0.0093	0.0098
5	0.0098	0.0098	0.0098	0.0078	0.0097	0.0097	0.0097

Table 5.3 Values of all-pairs power for [T1]–[B7] with $k = 4, n = 16, \alpha = 0.05$, and H^{A1}

Δ\ procedure	[T1]	[R2]	[T3]	[F4]	[H5]	[T6]	[B7]
3	0.0003	0.0026	0.0110	0.0110	0.0015	0.0505	0.0505
4	0.0114	0.0514	0.1294	0.1294	0.0364	0.2920	0.2920
5	0.1191	0.2753	0.4418	0.4418	0.2273	0.6372	0.6372

Table 5.4 Values of all-pairs power for [T1]–[B7] with $k = 4, n = 16, \alpha = 0.05$, and H^{A2}

Δ\ procedure	[T1]	[R2]	[T3]	[F4]	[H5]	[T6]	[B7]
3	0.0196	0.0460	0.0665	0.0832	0.0388	0.3187	0.3353
4	0.0846	0.1614	0.2143	0.2470	0.1407	0.5605	0.5990
5	0.2422	0.3810	0.4663	0.5045	0.3459	0.7817	0.8180

Table 5.5 Values of all-pairs power for [F1]–[B7] with $k = 5, n = 13, \alpha = 0.05$, and H^{A1}

Δ\ procedure	[T1]	[R2]	[T3]	[F4]	[H5]	[T6]	[B7]
4	0.0001	0.0020	0.0156	0.0156	0.0006	0.0744	0.0744
5	0.0050	0.0456	0.1565	0.1565	0.0196	0.3476	0.3476
6	0.0723	0.2489	0.4763	0.4763	0.1551	0.6780	0.6780

Table 5.6 Values of all-pairs power for [T1]–[B7] with $k = 5, n = 13, \alpha = 0.05$, and H^{A2}

Δ\ procedure	[T1]	[R2]	[T3]	[F4]	[H5]	[T6]	[B7]
4	0.0137	0.0390	0.0490	0.0590	0.0283	0.3810	0.3957
5	0.0621	0.1379	0.1635	0.1825	0.1072	0.5994	0.6294
6	0.1850	0.3318	0.3722	0.3973	0.2737	0.7915	0.8174

Table 5.7 Rate of reject of a pair for [T1]–[B7] with $k = 4, n = 16, \Delta = 3, \alpha = 0.05$, and H^{A1}

Pair\ procedure	[T1]	[R2]	[T3]	[F4]	[H5]	[T6]	[B7]
(1,2)	0.1790	0.2770	0.3001	0.3038	0.2498	0.4241	0.4303
(1,3)	0.7699	0.8398	0.8441	0.8435	0.8404	0.9097	0.9413
(1,4)	0.9915	0.9937	0.9937	0.9941	0.9962	0.9980	0.9997
(2,3)	0.1811	0.2784	0.3652	0.3744	0.2531	0.4940	0.5083
(2,4)	0.7701	0.8382	0.8430	0.8433	0.8389	0.9085	0.9409
(3,4)	0.1816	0.2797	0.3029	0.3058	0.2525	0.4265	0.4329

Table 5.8 Rate of reject of a pair for [T1]–[B7] with $k = 4$, $n = 16$, $\Delta = 3$, $\alpha = 0.05$, and H^{A2}

Pair\ procedure	[T1]	[R2]	[T3]	[F4]	[H5]	[T6]	[B7]
(1,2)	0.0106	0.0216	0.0217	0.0229	0.0103	0.0227	0.0246
(1,3)	0.1803	0.2158	0.2208	0.2374	0.2505	0.4256	0.4895
(1,4)	0.1799	0.2160	0.2212	0.2378	0.2508	0.5501	0.6726
(2,3)	0.1811	0.2180	0.2232	0.2397	0.2531	0.3187	0.3353
(2,4)	0.1820	0.2195	0.2246	0.2413	0.2542	0.4259	0.4910
(3,4)	0.0104	0.0211	0.0211	0.0226	0.0104	0.0222	0.0244

From Tables 5.3 and 5.5, the order of the power on H^{A1} is the following:

$$[B7] = [T6] > [F4] = [T3] > [R2] > [H5] > [T1].$$

Moreover, from Tables 5.4 and 5.6, the order of the power on H^{A2} is the following:

$$[B7] > [T6] > [F4] > [T3] > [R2] > [H5] > [T1].$$

[B7] and [T6] have high power because they assume simple order restrictions. Even though assuming them, [H5] is lower than [R2].

We also investigate the rate of reject for each pair. Tables 5.7 and 5.8 are the result of $(k, n) = (4, 16)$, $\Delta = 3$ and $\alpha = 0.05$.

The performance of power for a pair is similar to one of all-pairs power. However, in detail in Table 5.8, [H5] is better than [R2], [T3], and [F4] for H^{A2}. Furthermore, [H5] is slight better than [R2], [T3], and [F4] for far pairs in H^{A1}, but slight worse than [R2], [T3], and [F4] for near pairs in H^{A1} in Table 5.7.

5.4 Conclusion

All procedures well control the familywise error.

From the point of view of all-pairs power on alternative hypotheses and powers for each pair, we conclude that the procedure [B7] (as [3.4]) shows the best performance among all procedures if we assume simple order restrictions; otherwise, we should use the procedure [F4] (as [1.5]).

Reference

Ramsey PH (1978) Power differences between pairwise multiple comparisons. J Am Stat Assoc 73:479–485

Chapter 6
Application of Multiple Comparison Tests to Real Data

Abstract In this chapter, we apply multiple comparison tests presented in Chaps. 1–4 using real data. The procedures in this chapter are the seven procedures discussed in Chap. 5.

6.1 Introduction

This chapter presents the results of applying proposed procedures to real data.

We use the same notations as those in Chap. 5. Further, we use the seven procedures described in Chap. 5: [T1], [R2], [T3], [F4], [H5], [T6], and [B7]. [T1]–[F4] are all-pairwise comparisons (i.e., [1.1], [1.4], [1.3], and [1.5]) with no restriction (see Chap. 1) and [H5]–[B7] are all-pairwise comparisons (i.e., [3.1], [3.3], and [3.4]) with simple order restrictions (see Chap. 3).

6.2 Data

Table 6.1 is data of age-adjusted death rates in the Southern United States, collected from CDC (2015). The rate is the number of deaths per 100,000 total population. We get values about four major causes from the original table.

In point of view for homoscedasticity, we may use logarithmic transformed data for calculating estimates in the next section.

6.3 Critical Values

We dealt with this data in Shiraishi and Matsuda (2018) as the data with randomized blocks. Now we will apply the proposed procedures to this data as one-way layout.

T.-a. Shiraishi et al., *Pairwise Multiple Comparisons*,
JSS Research Series in Statistics,
https://doi.org/10.1007/978-981-15-0066-4_6

Table 6.1 Four major death rates in the Southern US in 2015

	Heart disease	Cancer	CLRD[a]	Accident
Delaware	165.2	165.6	42.5	46.0
Maryland	169.3	155.0	30.7	29.7
West Virginia	191.3	190.4	64.6	77.9
Virginia	154.2	159.5	37.1	39.6
North Carolina	162.4	164.7	45.5	47.9
South Carolina	177.8	166.6	49.9	54.0
Georgia	180.2	163.0	46.7	43.2
Florida	149.8	150.6	38.4	46.2
Kentucky	197.8	195.9	64.3	66.0
Tennessee	207.3	180.5	54.9	56.4
Alabama	229.7	175.6	56.4	50.9
Mississippi	240.5	188.4	57.1	59.8
Arkansas	223.2	185.4	62.4	49.6
Oklahoma	234.0	184.3	65.8	60.1
Louisiana	212.1	180.2	54.7	54.7
Texas	171.6	149.2	37.4	37.4

[a]Chronic lower respiratory diseases

Table 6.2 Critical values $F_m^{\ell-1}(\alpha(M,\ell))$ for the stepwise procedure [1.5] with $\alpha = 0.05$ and $m = 60$

$M \setminus \ell$	2	3	4	5	6	7	8	9	10
10	7.036	4.487	3.519	2.999	2.670	2.442	2.274	◊	2.040
9	6.827	4.367	3.430	2.926	2.608	2.387	◊	2.097	
8	6.594	4.234	3.332	2.846	2.538	◊	2.167		
7	6.332	4.083	3.220	2.754	◊	2.254			
6	6.033	3.911	3.092	◊	2.368				
5	5.683	3.708	◊	2.525					
4	5.261	◊	2.758						
3	◊	3.150							
2	4.001								

As $m = 60$ for this data, Tables 1.3 and 1.4 can be used for the critical values of the procedure [T1]–[T3] for $\alpha = 0.05, 0.01$, respectively. Moreover, we show tables for [F4] as Tables 6.2 and 6.3, tables for [H5] and [T6] as Tables 6.4 and 6.5, and tables for [B7] as Tables 6.6 and 6.7 for $\alpha = 0.05, 0.01$, respectively.

Table 6.3 Critical values $F_m^{\ell-1}(\alpha(M, \ell))$ for the stepwise procedure [1.5] with $\alpha = 0.01$ and $m = 60$

$M \setminus \ell$	2	3	4	5	6	7	8	9	10
10	10.435	6.405	4.923	4.138	3.648	3.310	3.061	◊	2.718
9	10.207	6.278	4.830	4.063	3.584	3.253	◊	2.823	
8	9.953	6.136	4.727	3.980	3.512	◊	2.953		
7	9.668	5.976	4.610	3.886	◊	3.119			
6	9.340	5.792	4.476	◊	3.339				
5	8.955	5.576	◊	3.649					
4	8.489	◊	4.126						
3	◊	4.977							
2	7.077								

Table 6.4 Critical values $td_1(\ell, m; \alpha(M, \ell))$ for the stepwise procedure [3.3] with $\alpha = 0.05$ and $m = 60$

$M \setminus \ell$	2	3	4	5	6	7	8	9	10
10	2.382	2.623	2.755	2.844	2.910	2.963	3.006	◊	3.074
9	2.339	2.582	2.714	2.803	2.870	2.922	◊	3.002	
8	2.291	2.536	2.668	2.758	2.824	◊	2.920		
7	2.236	2.482	2.616	2.705	◊	2.825			
6	2.171	2.420	2.554	◊	2.711				
5	2.093	2.344	◊	2.570					
4	1.995	◊	2.387						
3	◊	2.125							
2	1.671								

Table 6.5 Critical values $td_1(\ell, m; \alpha(M, \ell))$ for the stepwise procedure [3.3] with $\alpha = 0.01$ and $m = 60$

$M \setminus \ell$	2	3	4	5	6	7	8	9	10
10	2.992	3.215	3.339	3.424	3.488	3.540	3.583	◊	3.651
9	2.955	3.179	3.303	3.389	3.453	3.505	◊	3.584	
8	2.913	3.138	3.263	3.349	3.413	◊	3.508		
7	2.865	3.092	3.217	3.303	◊	3.420			
6	2.809	3.037	3.163	◊	3.315				
5	2.743	2.972	◊	3.186					
4	2.659	◊	3.020						
3	◊	2.785							
2	2.390								

Table 6.6 Critical values $\bar{b}_1^2(\ell, m; \alpha(M, \ell))$ for the stepwise procedure [3.4] with $\alpha = 0.05$ and $m = 60$

$M \setminus \ell$	2	3	4	5	6	7	8	9	10
10	5.673	6.362	6.652	6.795	6.867	6.898	6.904	$\diamond$	6.876
9	5.472	6.144	6.424	6.559	6.624	6.650	$\diamond$	6.639	
8	5.249	5.902	6.170	6.296	6.354	$\diamond$	6.372		
7	4.999	5.630	5.884	6.000	$\diamond$	6.064			
6	4.713	5.319	5.557	$\diamond$	5.702				
5	4.379	4.955	$\diamond$	5.265					
4	3.978	$\diamond$	4.712						
3	$\diamond$	3.963							
2	2.791								

Table 6.7 Critical values $\bar{b}_1^2(\ell, m; \alpha(M, \ell))$ for the stepwise procedure [3.4] with $\alpha = 0.01$ and $m = 60$

$M \setminus \ell$	2	3	4	5	6	7	8	9	10
10	8.953	9.887	10.335	10.595	10.759	10.868	10.940	$\diamond$	11.019
9	8.732	9.651	10.089	10.342	10.501	10.604	$\diamond$	10.717	
8	8.487	9.388	9.816	10.060	10.212	$\diamond$	10.374		
7	8.210	9.092	9.507	9.743	$\diamond$	9.979			
6	7.893	8.753	9.153	$\diamond$	9.514				
5	7.522	8.354	$\diamond$	8.949					
4	7.072	$\diamond$	8.232						
3	$\diamond$	7.256							
2	5.713								

6.4 Application to Real Data

Table 6.8 presents statistics $\max|T_{i'i}|$, which given (1.8), for [T1], [R2], and [T3], and $S(I)$, which given (1.40), for [F4], where we use the logarithmic transformed data from Table 6.1.

In [T1], we only use statistics $\max|T_{i'i}|$ for pair hypotheses in Table 6.8. Critical values for [T1] are 2.643 and 3.249 for $\alpha = 0.05, 0.01$, respectively. Thus, pair (1, 2), i.e., heart disease versus cancer, is retained for $\alpha = 0.05$. Moreover, pair (3, 4), i.e., chronic lower respiratory diseases versus accident, is also retained for $\alpha = 0.05$. The other pairs are rejected for $\alpha = 0.01$.

In [R2], we reach same result above. In detail, if we want to reject for $H(\{1, 2\})$ in [R2], we should reject all $H(\{1, 2\})$, $H(\{1, 2, 3\})$, $H(\{1, 2, 4\})$, and $H(\{1, 2, 3, 4\})$. For $\alpha = 0.05$, those evaluations are $1.501 < 2.294$, $20.337 > 2.403$, $20.137 > 2.403$, and $20.337 > 2.643$. Thus we retain $H(\{1, 2\})$ in [R2].

Table 6.8 Statistics $\max|T_{i'i}|$ and $S(I)$ for 4 major death rates

Column numbers	$\max \|T_{i'i}\|$	$S(I)$
(1, 2)	1.501	2.254
(1, 3)	20.337	413.597
(1, 4)	20.137	405.487
(2, 3)	18.836	354.781
(2, 4)	18.635	347.273
(3, 4)	0.200	0.040
(1, 2, 3)	20.337	256.877
(1, 2, 4)	20.137	251.671
(1, 3, 4)	20.337	273.041
(2, 3, 4)	18.836	234.031
(1, 2, 3, 4)	20.337	253.905

In [T3], we also reach same result above. In detail, if we want to reject for $H(\{1, 2\})$ in [T3], we should reject all $H(\{1, 2\})$, $H(\{1, 2, 3\})$, $H(\{1, 2, 4\})$, $H(\{1, 2\}, \{3, 4\})$, and $H(\{1, 2, 3, 4\})$. For $\alpha = 0.05$, those evaluations are $1.501 < 2.000, 20.337 > 2.403, 20.137 > 2.403, \max\{1.501, 0.200\} < 2.294$, and $20.337 > 2.643$. (We notice that the rejection of $H(\{1, 2\}, \{3, 4\})$ consists of satisfying either $|T_{12}| > 2.294$ or $|T_{34}| > 2.294$). Thus, we retain $H(\{1, 2\})$ in [T3].

We will explain the reason why [T3] differ from [R2] for the rejection of $H(\{1, 2\})$ above. First, in [T3], $H(\{1, 2\}, \{3, 4\})$ is added to the hypotheses of [R2]. The important point here is that the statistics of the pair (3, 4) may be larger than the critical value in order to rejecting $H(\{1, 2\}, \{3, 4\})$. If we clear the rejection of this hypothesis, the situation where all significance levels can be assigned to reject pair (1, 2) occurs, and the difference in critical values for $H(\{1, 2\})$ also occurs, i.e., 2.000 in [T3] and 2.294 in [R2]. If we can reject $H(\{1, 2\})$ in [R2], we can also reject $H(\{1, 2\}, \{3, 4\})$ in [T3] with the same critical value. Thus, there is no disadvantage to increasing $H(\{1, 2\}, \{3, 4\})$ in [T3]. Since it is also so for the other hypotheses, [T3] is uniformly more powerful than [R2].

In [F4], we also reach same result above. In detail, if we want to reject for $H(\{1, 2\})$ in [F4], we should reject all $H(\{1, 2\})$, $H(\{1, 2, 3\})$, $H(\{1, 2, 4\})$, $H(\{1, 2\}, \{3, 4\})$, and $H(\{1, 2, 3, 4\})$. For $\alpha = 0.05$, those evaluations are $2.254 < 4.001$, $256.877 > 3.150$, $251.671 > 3.150$, $\max\{2.254, 0.040\} < 5.261$, and $253.905 > 2.758$. (We notice that the rejection of $H(\{1, 2\}, \{3, 4\})$ consists of satisfying either $S(12) > 5.261$ or $S(34) > 5.261$). Thus, we retain $H(\{1, 2\})$ in [F4].

Next, we will apply the procedures with simple order restrictions. In this case, we use the opposite order against Chap. 3, because four major causes of death are listed in descending order of death rates across the United States.

Table 6.9 presents statistics $-T_{i'i}$ and $\max(-T_{i'i})$, which given (1.8), for [H5] and [T6], respectively, and $\bar{B}_1^2(I)$, which given (3.49), for [B7].

Table 6.9 Statistics $-T_{i'i}$, $\max(-T_{i'i})$ and $\bar{B}_1^2(I)$ for 4 major death rates

Column numbers	$-T_{i'i}$	$\max(-T_{i'i})$	$\bar{B}_1^2(I)$
(1, 2)	1.501	1.501	2.254
(1, 3)	20.337	–	–
(1, 4)	20.137	–	–
(2, 3)	18.836	18.836	354.781
(2, 4)	18.635	–	–
(3, 4)	−0.200	−0.200	0
(1, 2, 3)	–	20.337	513.755
(2, 3, 4)	–	18.836	468.022
(1, 2, 3, 4)	–	20.337	761.676

In [H5], we use statistics $-T_{i'i}$ for pair hypotheses in Table 6.9. Critical values for [H5] are 2.387 and 3.020 for $\alpha = 0.05, 0.01$, respectively. Thus pair (1, 2), i.e., heart disease versus cancer, is retained for $\alpha = 0.05$. Moreover, pair (3, 4), i.e., chronic lower respiratory diseases versus accident, is also retained for $\alpha = 0.05$. The other pairs are rejected for $\alpha = 0.01$.

In [T6], we assume the simple order restriction by using the order of result of United States. In the Southern United States, however, order of chronic lower respiratory diseases versus accident is exchanged. Therefore, the value of $-T_{i'i}$ of $H(\{3, 4\})$ becomes −0.200. Regardless of this, the test result of [T6] is same as [H5], i.e., pairs (1, 2) and (3, 4) are retained for $\alpha = 0.05$. In detail, if we want to reject for $H(\{1, 2\})$ in [T6], we should reject all $H(\{1, 2\})$, $H(\{1, 2, 3\})$, $H(\{1, 2\}, \{3, 4\})$, and $H(\{1, 2, 3, 4\})$. For $\alpha = 0.05$, those evaluations are $-0.200 < 1.671$, $20.337 > 2.125$, $\max\{1.501, -0.200\} < 1.995$, and $20.337 > 2.387$. Thus, we retain $H(\{1, 2\})$ in [T6].

In [B7], we assume the same condition above. Therefore, the value of $\bar{B}_1^2(I)$ of $H(\{3, 4\})$ becomes 0. Regardless of this, the test result of [B7] is same as [T6], i.e., pairs (1, 2) and (3, 4) are retained for $\alpha = 0.05$. In detail, if we want to reject for $H(\{1, 2\})$ in [B7], we should reject all $H(\{1, 2\})$, $H(\{1, 2, 3\})$, $H(\{1, 2\}, \{3, 4\})$, and $H(\{1, 2, 3, 4\})$. For $\alpha = 0.05$, those evaluations are $0 < 2.791$, $513.755 > 3.963$, $\max\{2.254, 0\} < 3.978$, and $761.676 > 4.712$. Thus, we retain $H(\{1, 2\})$ in [B7].

6.5 Discussion

We will compare the result of the previous section with one of Shiraishi and Matsuda (2018).

Table 6.10 shows all results for this comparison.

In Shiraishi and Matsuda (2018), [B6]* is an analogous procedure to [B7]. [B6]* uses Table 6.1 as the data in which states are randomized blocks. Only this procedure detects the difference of pair $(1, 2)$. This means that it is important to choose a suitable model.

Table 6.10 All results of proposed procedures for four major death rates ($\alpha = 0.05$)

Column numbers	[T1]	[R2]	[T3]	[F4]	[H5]	[T6]	[B7]	[B6]*
(1, 2)	Retain	Retain	Retain	Retain	Retain	Retain	Retain	Reject
(1, 3)	Reject	Reject	Reject	Reject	Reject	Reject	Reject	Reject
(1, 4)	Reject	Reject	Reject	Reject	Reject	Reject	Reject	Reject
(2, 3)	Reject	Reject	Reject	Reject	Reject	Reject	Reject	Reject
(2, 4)	Reject	Reject	Reject	Reject	Reject	Reject	Reject	Reject
(3, 4)	Retain	Retain	Retain	Retain	Retain	Retain	Retain	Retain

[B6]* is the procedure in Shiraishi and Matsuda (2018)

References

Centers for disease control and prevention (2015) National Center for Health Statistics. https://www.cdc.gov/nchs/pressroom/stats_of_the_states.htm

Shiraishi T, Matsuda S (2018) Closed testing procedures for all pairwise comparisons in a randomized block design. Commun Stat Theory Methods 47:3571–3587

Chapter 7
Computation of Distribution Functions for Statistics Under Simple Order Restrictions

Abstract To execute the multiple comparison methods introduced in Chaps. 1–4, we need the upper $100\alpha\%$ points for the distributions of the statistics. For cases with no order constraints on k means (see Chaps. 1, 2, and 4), we can use executable programs listed in Appendix B of Shiraishi (Multiple comparison procedures under continuous distributions. Kyoritsu Shuppan Co., Ltd., 2011) for the values of the upper $100\alpha\%$ points of the distributions. On the other hand, when there are order constraints on the location parameters introduced in Chap. 3, the multiple comparison methods require more complicated numerical calculations for the statistical distribution. The distributions in these cases are derived from the normal distribution. The key objective of the numerical calculation is the integral transformation of the Gaussian function. We set the function family **G** as a generalization of the Gaussian function and introduce Sinc approximation method suitable for the approximation and calculus of functions belonging to **G**. As a specific application, in Sect. 7.2, we introduce calculation methods concerning the distribution function of the Hayter type statistics, which is the distribution of the max statistics in Chap. 3. Then, in Sect. 7.3, we introduce a numerical method to calculate the level probability for the multiple comparison methods that is based on the sum-of-squares statistics described in Sect. 3.3. The proofs for the theorems are not presented in this chapter. Please visit the following webpage for the proofs: http://www.st.nanzan-u.ac.jp/info/sugiurah/sincstatistics.

7.1 Function Family G and Sinc Approximation

Sinc approximation is a method suitable for the approximation and calculus of an analytic function $f(x)$ $(x \in \mathbb{R})$ rapidly decaying to zero as $|x| \to \infty$. In particular, when $f(x)$ belongs to the function family **G** described in Sect. 7.1.1, the approximation is highly accurate.

Family **G** is closed with respect to linear combination, product, shift of variables, differentiation, and convolution, among others. It contains density functions of various distributions derived from the normal distribution. These density functions can be approximated using Sinc interpolation, and the corresponding distribution function can be accurately approximated by Sinc integral.

T.-a. Shiraishi et al., *Pairwise Multiple Comparisons*,
JSS Research Series in Statistics,
https://doi.org/10.1007/978-981-15-0066-4_7

Furthermore, the application of Sinc approximation is simple. The approximation function is given as a series expansion on a Sinc basis. The expansion coefficients are the sample values themselves on the equally spaced sample points and require no cost except the evaluation of the sample values.

7.1.1 Function Family **G**

In this section, we introduce the function family **G**. In the following, we express complex numbers and complex variables in Greek letters, their real part, and imaginary part with corresponding Roman letters and the same letters putted primes. For example, we write a complex number $\xi \in \mathbb{C}$ as $\xi = x + ix'$ where $x, x' \in \mathbb{R}$. Specifically, $\alpha,\ \beta$ are used to represent positive real parameters.

The Gauss function $G(x) = e^{-x^2}$ can be continued analytically to *the entire function* $G(\xi)$ which is analytic in whole complex plane $\mathbb{C}$, and satisfies

$$|G(\xi)| = \left|e^{-(x+ix')^2}\right| = e^{-x^2+x'^2} \qquad (\xi = x + ix' \in \mathbb{C}).$$

For the generalization of $G(x)$, we consider a function $f(x) \in \mathbb{C}$ for $x \in \mathbb{R}$ which can be continued analytically to the entire function on $\mathbb{C}$ and satisfies

$$|f(\xi)| \le Ae^{-\alpha x^2+\beta x'^2} \qquad (\xi = x + ix' \in \mathbb{C}) \tag{7.1}$$

for some positive numbers $A, \alpha, \beta > 0$. We define the function family **G** of the set of all these functions (Shiraishi and Sugiura 2018). From Theorem 7.7 described later, one can see that $\alpha \le \beta$ is sufficient.

We define subfamilies $\mathbf{G}(\alpha, \beta)$, $\mathbf{G}(A, \alpha, \beta)$ of **G** as

$$\mathbf{G}(\alpha, \beta) = \left\{ f \in \mathbf{G} \;\middle|\; \sup_{x,x'\in\mathbb{R}} \frac{|f(x+ix')|}{e^{-\alpha x^2+\beta x'^2}} < \infty \right\}, \quad \alpha, \beta > 0,$$

$$\mathbf{G}(A, \alpha, \beta) = \left\{ f \in \mathbf{G} \;\middle|\; |f(x+ix')| \le Ae^{-\alpha x^2+\beta x'^2} \right\}, \quad A, \alpha, \beta > 0.$$

We can see that

$$\mathbf{G}(\alpha, \beta) = \bigcup_{A>0} \mathbf{G}(A, \alpha, \beta),$$

$$\mathbf{G} = \bigcup_{\alpha>0,\beta>0} \mathbf{G}(\alpha, \beta) = \bigcup_{A>0,\alpha>0,\beta>0} \mathbf{G}(A, \alpha, \beta).$$

We call A, α, β *the Gauss parameters* of a function $f \in \mathbf{G}(A, \alpha, \beta)$.

We show some properties of **G**.

Theorem 7.1 $\mathbf{G}$ *is closed with respect to linear combination. That is, if* $f_1 \in \mathbf{G}(A_1, \alpha_1, \beta_1)$, $f_2 \in \mathbf{G}(A_2, \alpha_2, \beta_2)$ *then, for arbitrary* $a_1, a_2 \in \mathbb{C}$,

$$a_1 f_1 + a_2 f_2 \in \mathbf{G}\left(|a_1|\, A_1 + |a_2|\, A_2, \min\{\alpha_1, \alpha_2\}, \max\{\beta_1, \beta_2\}\right).$$

Theorem 7.2 $\mathbf{G}$ *is closed with respect to product. That is, if* $f_1 \in \mathbf{G}(A_1, \alpha_1, \beta_1)$, $f_2 \in \mathbf{G}(A_2, \alpha_2, \beta_2)$ *then*

$$f_1 f_2 \in \mathbf{G}\left(A_1 A_2, \alpha_1 + \alpha_2, \beta_1 + \beta_2\right).$$

Theorem 7.3 $\mathbf{G}$ *is closed with respect to linear translation of variable. That is, let* $g(x) = f(ax + \gamma)$ *for* $f \in \mathbf{G}(\alpha, \beta)$, $a \in \mathbb{R}\backslash\{0\}$ *and* $\gamma \in \mathbb{C}$, *then*

$$g \in \mathbf{G}(\alpha', \beta')$$

is valid for arbitrary α', β' *such that* $0 < \alpha' < a^2\alpha$ *and* $\beta' > a^2\beta$.

Theorem 7.4 $\mathbf{G}$ *is closed with respect to differentiation. That is, if* $f \in \mathbf{G}(\alpha, \beta)$ *then*

$$f' \in \mathbf{G}(\alpha', \beta')$$

for arbitrary α', β' *such that* $0 < \alpha' < \alpha$ *and* $\beta' > \beta$.

We denote *the indefinite integral* F of a function f by

$$F(x) = \int_{-\infty}^{x} f(y)dy,$$

with the lower limit of the integral $-\infty$, unless otherwise stated. For $f \in \mathbf{G}$, the integral F is continued analytically to the entire function

$$F(\xi) = \int_{-\infty}^{x} f(y + ix')dy \quad (\xi = x + ix' \in \mathbb{C}).$$

Theorem 7.5 *Let* $f_1 \in \mathbf{G}(A_1, \alpha_1, \beta_1)$, $f_2 \in \mathbf{G}(A_2, \alpha_2, \beta_2)$. *Then the indefinite integral* F_1 *of* f_1 *satisfies*

$$F_1 f_2 \in \mathbf{G}\left(\sqrt{\frac{\pi}{\alpha_1}}\, A_1 A_2, \alpha_2, \beta_1 + \beta_2\right).$$

We denote *the Fourier transform* of $f(x)$ by

$$\hat{f}(y) = \frac{1}{\sqrt{2\pi}} \int_{-\infty}^{\infty} f(x)e^{ixy}dx.$$

The analytic continuation $\hat{f}(\eta)$ on $\mathbb{C}$ is

$$\hat{f}(\eta) = \frac{1}{\sqrt{2\pi}} \int_{-\infty}^{\infty} f(x) e^{ix\eta} dx \quad (\eta \in \mathbb{C}).$$

The inverse Fourier transform of $\hat{f}(y)$ is

$$f(x) = \frac{1}{\sqrt{2\pi}} \int_{-\infty}^{\infty} \hat{f}(y) e^{-ixy} dy.$$

and its analytic continuation of $f(x)$ on $\mathbb{C}$ is

$$f(\xi) = \frac{1}{\sqrt{2\pi}} \int_{-\infty}^{\infty} \hat{f}(y) e^{-i\xi y} dy \quad (\xi \in \mathbb{C}).$$

We have the following theorem for Fourier transform.

Theorem 7.6 *$\mathbf{G}$ is closed with respect to the Fourier transform. That is,*

$$f \in \mathbf{G}(A, \alpha, \beta) \quad \Rightarrow \quad \hat{f} \in \mathbf{G}\left(\tfrac{A}{\sqrt{2\alpha}}, \tfrac{1}{4\beta}, \tfrac{1}{4\alpha}\right),$$
$$\hat{f} \in \mathbf{G}(A, \alpha, \beta) \quad \Rightarrow \quad f \in \mathbf{G}\left(\tfrac{A}{\sqrt{2\alpha}}, \tfrac{1}{4\beta}, \tfrac{1}{4\alpha}\right).$$

Theorem 7.6 deduces the interesting property below.

Theorem 7.7 *If $G(\alpha, \beta) \neq \{0\}$ then $\alpha \leq \beta$.*

We denote *the convolution* of functions $f,\ g$ on $\mathbb{R}$ by

$$f * g(x) = \int_{-\infty}^{\infty} f(y - x) g(y) dy.$$

For $f \in \mathbf{G}$, the convolution $f * g$ is continued analytically to the entire function

$$f * g(\xi) = \int_{-\infty}^{\infty} f(y - \xi) g(y) dy \quad (\xi \in \mathbb{C}).$$

We have the following theorem for convolution.

Theorem 7.8 *Let $f \in \mathbf{G}(A, \alpha, \beta)$ and $g \in \mathbf{G}(A', \alpha', \beta')$ then*

$$f * g \in \mathbf{G}\left(AA'\sqrt{\frac{\pi\beta\beta'}{\alpha\alpha'(\beta + \beta')}}, \frac{\alpha\alpha'}{\alpha + \alpha'}, \frac{\beta\beta'}{\beta + \beta'}\right).$$

Theorem 7.9 *Let $f \in \mathbf{G}(A, \alpha, \beta)$ and $g(x)$ $(x \in \mathbb{R})$ be an integrable function satisfying*

$$|g(x)| \leq A' e^{-\alpha' x^2} \quad (x \in \mathbb{R}).$$

Then

$$f * g \in \mathbf{G}\left(AA' \sqrt{\frac{\pi}{\alpha + \alpha'}},\ \frac{\alpha \alpha'}{\alpha + \alpha'},\ \beta \right).$$

As described above, function family **G** is closed with respect to some important operations. It provides a good foundation when we characterize functions derived from the Gauss function and analyze them.

7.1.2 Sinc Approximation

We describe Sinc approximation for *a rapidly decreasing function* $f(x)$ on $x \in \mathbb{R}$. We say $f(x)$ is rapidly decreasing when $|f(x)| = o(|x|^{-\ell}) \quad (|x| \to \infty)$ for every positive integer $\ell > 0$.

The Sinc function

$$\text{sinc}(x) := \begin{cases} \dfrac{\sin x}{x} & (x \neq 0), \\ 1 & (x = 0), \end{cases} \tag{7.2}$$

is a real function on $x \in (-\infty, \infty)$ which can be continued analytically to the entire function. It satisfies $|\text{sinc}(x)| \leq 1 \quad (x \in \mathbb{R})$.

We take an equidistant sample points

$$\boldsymbol{x} = (x_k), \quad x_k = x_0 + kh \ (k \in \mathbb{Z}),$$

with *the sample origin* x_0 and *the sample spacing* $h > 0$. Let

$$\boldsymbol{f} = (f_k), \quad f_k = f(x_k) \ (k \in \mathbb{Z}),$$

be function values $f(x)$ on $\boldsymbol{x}$. We define *Sinc basis*

$$s_h(x_k, x) := \text{sinc}\left(\frac{\pi(x - x_k)}{h} \right) \qquad (k \in \mathbb{Z}), \tag{7.3}$$

on $\boldsymbol{x}$ (Fig. 7.1) with the property

$$|s_h(x_k, x)| \leq 1 \qquad (x \in \mathbb{R}, \ \ k \in \mathbb{Z}). \tag{7.4}$$

On $x_j \ (j \in \mathbb{Z})$, the values of the Sinc basis $s_h(x_k, x)$ are

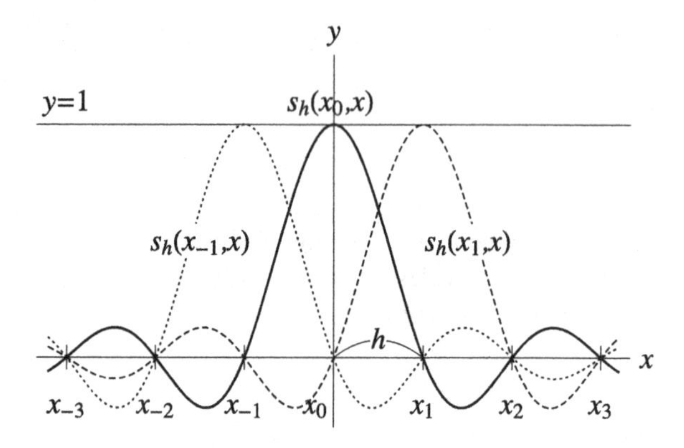

Fig. 7.1 Sinc basis

$$s_h(x_k, x_j) = \text{sinc}\,(\pi(j-k)) := \delta_{jk} = \begin{cases} 1 \ (j = k), \\ 0 \ (j \neq k), \end{cases} \tag{7.5}$$

where δ_{jk} denotes the Kronecker delta.

Sinc interpolation is the series expansion of $f(x)$ on a Sinc basis

$$c_x[f](x) := \sum_{k=-\infty}^{\infty} f_k s_h(x_k, x) \cong f(x) \qquad (x \in \mathbb{R}), \tag{7.6}$$

with the coefficients $f_k = f(x_k),\ k \in \mathbb{Z}$. Here, $\cong$ represents that the left-hand side is approximately equal to the right-hand side. From (7.5), we see that the Sinc interpolation satisfies the interpolation condition,

$$c_x[f](x_j) = \sum_{k=-\infty}^{\infty} f_k \delta_{jk} = f(x_j) \qquad (j \in \mathbb{Z}). \tag{7.7}$$

That is, $c_x[f](x)$ is a function interpolating $f(x)$ on the sampling points.

For the indefinite integral

$$F(x) = \int_{-\infty}^{x} f(t)dt$$

of $f(x)$, *Sinc integral*

$$C_x[f](x) := \int_{-\infty}^{x} c_x[f](t)dt = \sum_{k=-\infty}^{\infty} f_k S_h(x_k, x) \cong F(x), \tag{7.8}$$

is an approximation to $F(x)$. Here,

$$S_h(x_k, x) := \int_{-\infty}^{x} s_h(x_k, t)dt = \frac{h}{\pi}\left\{\mathrm{Si}\left(\frac{\pi(x - x_k)}{h}\right) + \frac{\pi}{2}\right\}, \tag{7.9}$$
$$\mathrm{Si}(x) := \int_0^x \frac{\sin t}{t}dt, \quad \mathrm{Si}(\pm\infty) = \pm\frac{\pi}{2},$$

with *the sine integral* $\mathrm{Si}(x)$, a standard function in several computing systems such as Mathematica.

For $|\mathrm{Si}(x)| < 2$ $(x \in \mathbb{R})$, it holds that

$$|S_h(x_k, x)| < \frac{4h}{\pi} \qquad (x \in \mathbb{R}). \tag{7.10}$$

We approximate

$$Q[f] := \int_{-\infty}^{\infty} f(t)dt = F(\infty)$$

by $C_x[f](\infty) \cong F(\infty)$. Since from (7.9), we see that

$$S_h(x_k, \infty) = \frac{h}{\pi}\left\{\mathrm{Si}\,(\infty) + \frac{\pi}{2}\right\} = h,$$

we have

$$C_x[f](\infty) = h \sum_{k=-\infty}^{\infty} f_k.$$

We denote this by

$$T_x[f] := h \sum_{k=-\infty}^{\infty} f_k \cong Q[f], \tag{7.11}$$

and call it *the trapezoidal rule*.

Since $f(x)$ is rapidly decreasing as $|x| \to \infty$, $|f_k| = |f(x_k)|$ is negligibly small when $|k|$ is large. So, in the actual calculation, the truncated finite series of (7.6), (7.8), and (7.11)

$$c_x^n[f](x) := \sum_{k=-n}^{n} f_k s_h(x_k, x) \cong c_x[f](x),$$
$$C_x^n[f](x) := \sum_{k=-n}^{n} f_k S_h(x_k, x) \cong C_x[f](x), \tag{7.12}$$
$$T_x^n[f] := h \sum_{k=-n}^{n} f_k \qquad\qquad \cong T_x[f],$$

are sufficient. Here the number of sample points N is equal to $2n + 1$. For the three equations in (7.12), we call them *the finite Sinc interpolation*, *the finite Sinc integral*, and *the finite trapezoidal rule*, respectively. We call them all together *Sinc approximation*.

From (7.5), we see that the finite Sinc interpolation satisfies the interpolation condition

$$c_{\boldsymbol{x}}^{n}[f](x_j) = \sum_{k=-n}^{n} f_k \delta_{jk} = f(x_j) \qquad (-n \le j \le n). \tag{7.13}$$

The smaller the number of sample points N is, the more efficient the calculation is. From the inequalities (7.4), (7.10), we have

$$|f_k s_h(x_k, x)| \le |f_k|,$$
$$|f_k S_h(x_k, x)| < \frac{4h}{\pi}|f_k|.$$

Therefore, the series terms on the right-hand side of (7.12) can be ignored if $|f_k|$ is small.

For a function $f(x)$ and a given tolerance $\varepsilon > 0$, we set an interval $I_\varepsilon = [a_\varepsilon, b_\varepsilon]$ satisfying

$$|f(x)| < \varepsilon \quad (x \notin I_\varepsilon).$$

We call this *an approximate support* for the tolerance ε, or simply *ε support.*

Let $c_\varepsilon = (a_\varepsilon + b_\varepsilon)/2$ and $R_\varepsilon = (b_\varepsilon - a_\varepsilon)/2$ be the midpoint and the half width of I_ε respectively. Then, we set sample points $\boldsymbol{x} = (x_k),\ x_k = x_0 + kh$ $(k \in \mathbb{Z})$ with

$$x_0 = c_\varepsilon,\ h = \lfloor R_\varepsilon/n \rfloor,$$

where $\lfloor\cdot\rfloor$ is *the floor function* $\lfloor\cdot\rfloor$ and $\lfloor x \rfloor$ is the greatest integer less than or equal to the real number x. In this setting, if $|k| > n$ then $x_k \notin I_\varepsilon$ and $|f_k| = |f(x_k)| < \varepsilon$. Hence, it is possible to neglect small terms in the right-hand side series of (7.12).

7.1.3 Numerical Examples

In this section, we give numerical examples to demonstrate the performance of finite Sinc approximation. We approximate the density $\varphi(x)$, the distribution $\Phi(x)$, and the total probability $\Phi(\infty) = 1$ of the standard normal distribution by finite Sinc interpolation, finite Sinc integration, and the finite trapezoidal rule, respectively. Numerical computations are done with the double-precision arithmetic having about 16 significant digits.

Let $\varepsilon = 10^{-14}$ be a tolerance and set an approximate support $I_\varepsilon = [-R_\varepsilon, R_\varepsilon]$, $R_\varepsilon = 8$, for $\varphi(x)$ (Fig. 7.2), because we have

$$\varphi(x) < \varphi(R_\varepsilon) \cong 5 \times 10^{-15} \qquad (|x| > R_\varepsilon).$$

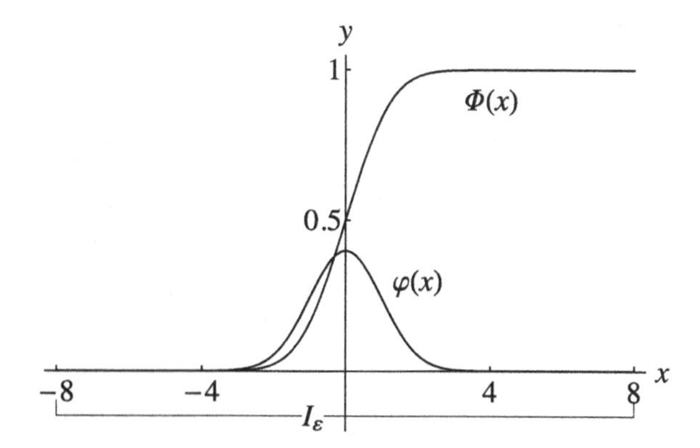

Fig. 7.2 The approximation support $I_\varepsilon,\ \varepsilon = 10^{-14}$

Let n be a natural number. We set the sample spacing $h = R_\varepsilon/n$ and the sample points $\boldsymbol{x} = (kh),\ k \in \mathbb{Z}$. The magnitude of samples $\varphi(kh)$ is rapidly and monotonically decreasing to zero as $|k| \to \infty$. Hence, $|\varphi(kh)| \ < \varepsilon\ (|k| > n)$ is satisfied for this setting.

In this setting, we evaluate $N = 2n + 1$ sample values $\varphi_k = \varphi(kh),\ -n \leq k \leq n$, and calculate the finite Sinc approximations

$$
\begin{aligned}
f_n(x) = c_{\boldsymbol{x}}^n[\varphi](x) &:= \sum_{k=-n}^{n} \varphi_k s_h(kh, x) \cong \varphi(x), \\
F_n(x) = C_{\boldsymbol{x}}^n[\varphi](x) &:= \sum_{k=-n}^{n} \varphi_k S_h(kh, x) \cong \Phi(x), \\
Q_n = T_{\boldsymbol{x}}^n[\varphi] &:= h \sum_{k=-n}^{n} \varphi_k \qquad\qquad \cong \Phi(\infty) = 1.
\end{aligned}
\tag{7.14}
$$

First, we show the graph of the errors

$$
e_n(x) = f_n(x) - \varphi(x), \quad E_n(x) = F_n(x) - \Phi(x)
$$

with $n = 7$ ($N = 15$, the number of samples) in Figs. 7.3 and 7.4.

These figures show that $f_n(x)$ and $F_n(x)$ is sufficiently accurate not only on $I_\varepsilon = [-8, 8]$ but also its outside. We will see in Sect. 7.1.4 that while we sample $f(x)$ on the approximate support I_ε, the Sinc approximations $f_n(x)$ and $F_n(x)$ give uniform approximations on the infinite interval.

From Figs. 7.3 and 7.4, the maximum absolute errors are given by

$$
\begin{aligned}
e_n &:= \max_{-20 \leq x \leq 20} |e_n(x)| < 2.5 \times 10^{-3}, \\
E_n &:= \max_{-20 \leq x \leq 20} |E_n(x)| < 10^{-3}.
\end{aligned}
$$

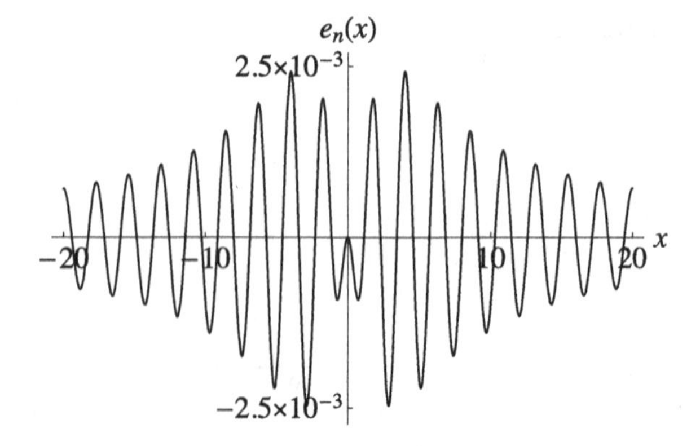

Fig. 7.3 The error of the finite Sinc interpolation ($n = 7$)

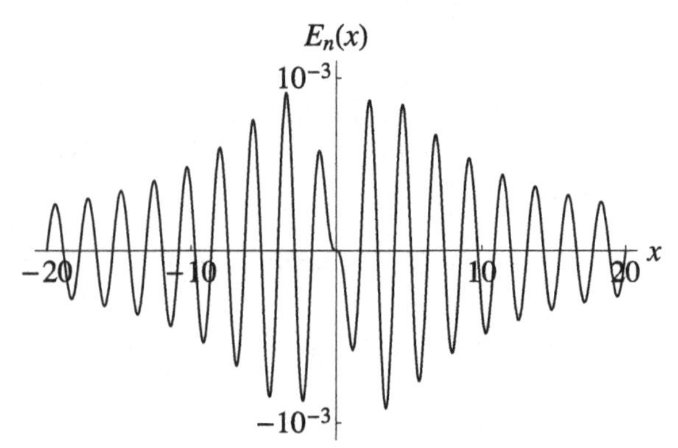

Fig. 7.4 The error of the finite Sinc integral ($n = 7$)

The absolute error of the finite trapezoidal rule, $|Q_n - \Phi(\infty)| = 5.5 \times 10^{-7}$, is much smaller than the errors of the finite Sinc interpolation and the finite Sinc integration. We will explain this fact in Sect. 7.1.4.

For comparison, we consider the natural cubic spline which can handle function approximation and indefinite integration in a unified manner.

We construct the natural cubic spline approximation to $\varphi(x)$,

$$g_n(x) \cong \varphi(x) \quad (x \in I_\varepsilon),$$

on the same sample points $kh, \ -n \le k \le n$. Since $g_n(x)$ is a piecewise polynomial, we can easily integrate it and get an approximation to $\Phi(x)$,

$$G_n(x) := \int_{-8}^{x} g_n(x)dx \cong \int_{-\infty}^{x} \varphi(x)dx = \Phi(x) \quad (x \in I_\varepsilon).$$

For $n = 7$, we show the graphs of the errors

$$e_n(x) = g_n(x) - \varphi(x), \quad E_n(x) = G_n(x) - \Phi(x)$$

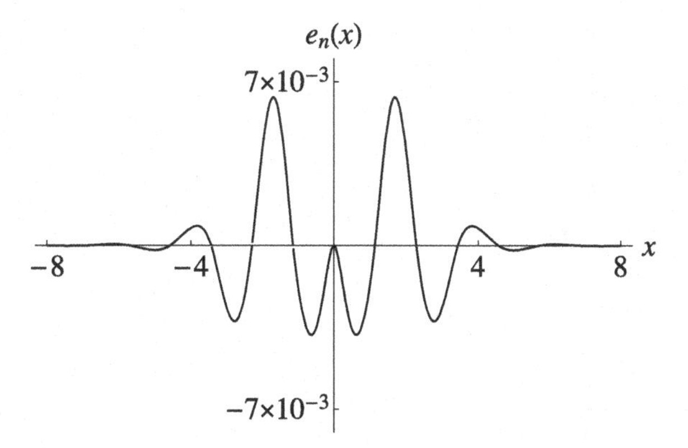

Fig. 7.5 The error of the spline interpolation ($n = 7$)

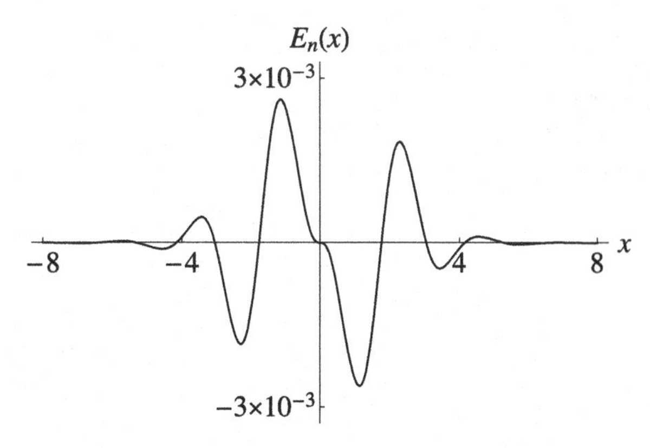

Fig. 7.6 The error of the spline integral ($n = 7$)

in Figs. 7.5 and 7.6. The range of variable $x \in I_\varepsilon = [-8, 8]$ is the domain of the spline function. The spline approximations guarantee no accuracy outside I_ε.

From these figures, the maximum absolute errors are given by

$$e_n := \max_{-8\leq x\leq 8} |e_n(x)| < 7 \times 10^{-3},$$
$$E_n := \max_{-8\leq x\leq 8} |E_n(x)| < 3 \times 10^{-3}.$$

We see that for $n = 7$ the finite Sinc approximations have almost the same accuracy as the natural cubic spline approximations.

Next, we examine the accuracy for each value n. Figures 7.7, 7.8 show the maximum absolute errors e_n, E_n for $e_n(x)$, $E_n(x)$ $(1 \leq n \leq 30)$ on $x \in I_\varepsilon$. The solid curve represents the finite Sinc approximations, the dotted curve the cubic natural spline approximations.

Figures 7.7 and 7.8 suggest that the errors e_n and E_n for the finite Sinc approximations decrease faster than the order $O(r^n)$ for any $0 < r < 1$. Since the geometric sequence of n is drawn on a straight line in a semilogarithmic graph, we can see that e_n, E_n of the finite Sinc approximations are decreasing faster than the geometric sequence.

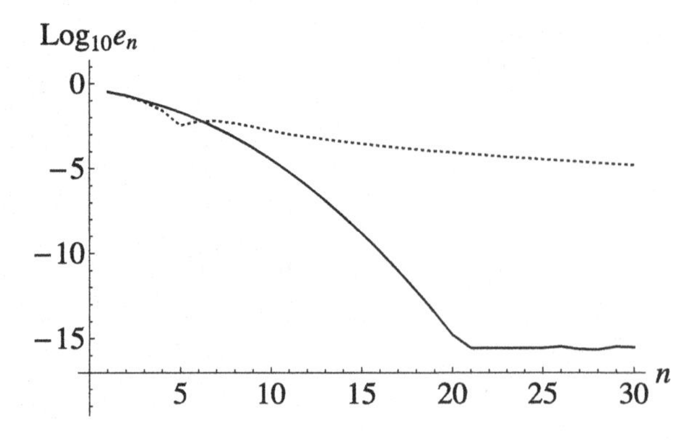

Fig. 7.7 The maximum absolute error of the finite Sinc interpolation and the spline interpolation

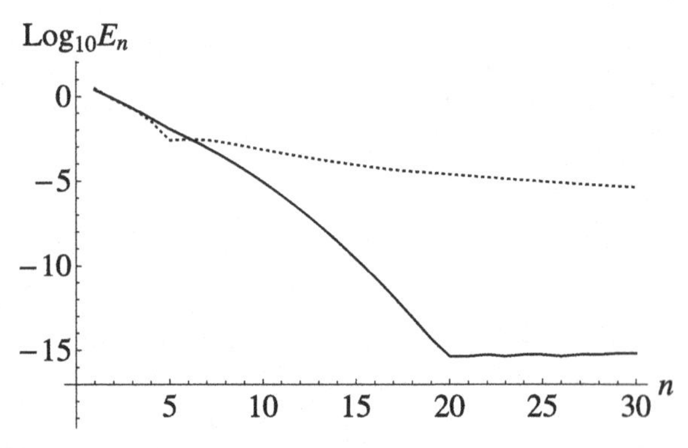

Fig. 7.8 The maximum absolute errors of the finite Sinc integral and the spline integral

The errors e_n, E_n for the Sinc approximations decrease to the rounding error level 10^{-16} of double precision and not decrease anymore as n grows.

In Sect. 7.1.4, we show that when we can ignore the truncation errors, there exist a positive constant $0 < r < 1$ and the maximum absolute errors of finite Sinc interpolation and finite Sinc integration satisfies

$$e_n = O(n^{-1}r^{n^2}), \quad E_n = O(n^{-2}r^{n^2}). \tag{7.15}$$

On the other hand, for the natural cubic spline approximations, $e_n = O(n^{-4})$ and $E_n = O(n^{-4})$. Finite Sinc approximations are significantly superior for large n to natural spline approximations.

It is advisable to choose cubic natural spline approximations for obtaining the approximations of $2 \sim 3$ significant digits and finite Sinc approximations for those of more than four significant digits.

Finally, in Fig. 7.9, we show the graph of the absolute error $|Q_n - 1|$ for the finite trapezoidal rule to approximate $\Phi(\infty) = 1$. The error decreases much more faster than those of the finite Sinc interpolation and the finite Sinc integration.

In fact, In Sect. 7.1.4, we show that $|Q_n - 1| = O\left(n^{-1}r^{(2n)^2}\right)$ with the same r in (7.15).

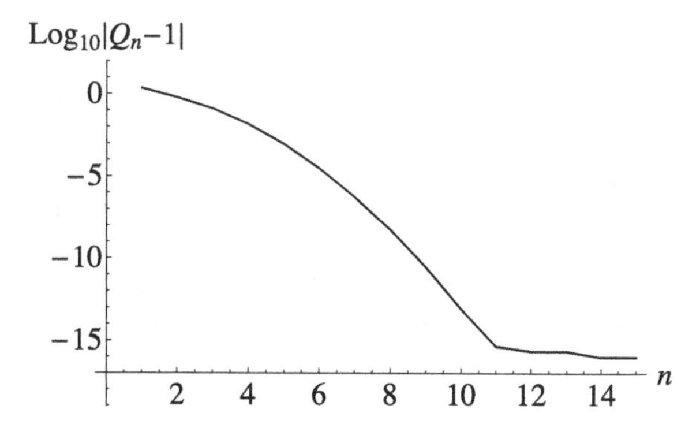

Fig. 7.9 The absolute error of the finite trapezoidal rule

7.1.4 Error Analysis for Finite Sinc Approximation

In this section, we describe the error analysis for finite Sinc approximation on the function family **G**. For general theory of Sinc approximation, see Stenger (1993). Also, in Lund and Bowers (1992), various applications are introduced.

Let $f \in \mathbf{G}(A, \alpha, \beta)$. Then,

$$|f(\xi)| \le Ae^{-\alpha x^2+\beta x'^2} \qquad (\xi = x + ix' \in \mathbb{C}). \tag{7.16}$$

Let an interval $I = [-R, R],\ R > 0,$ be an approximate support for $f(x)$. For a natural number n, we take sample points $\boldsymbol{x} = (kh), k \in \mathbb{Z}$ with the sample spacing $h = R/n$ and investigate the errors of the finite Sinc approximations.

We denote the errors of the Sinc approximations as

$$\begin{aligned} Ec_{\boldsymbol{x}}[f](x) &:= c_{\boldsymbol{x}}[f](x) - f(x), \\ EC_{\boldsymbol{x}}[f](x) &:= C_{\boldsymbol{x}}[f](x) - F(x), \\ ET_{\boldsymbol{x}}[f] &:= T_{\boldsymbol{x}}[f] - I[f], \end{aligned}$$

and call them *the discretization errors.*

We denote the differences of the finite Sinc approximations for the Sinc approximations as

$$\begin{aligned} \check{E}c_{\boldsymbol{x}}^n[f](x) &:= c_{\boldsymbol{x}}^n[f](x) - c_{\boldsymbol{x}}[f](x) = -\sum_{|k|>n} f(kh)s_h(x_k, x), \\ \check{E}C_{\boldsymbol{x}}^n[f](x) &:= C_{\boldsymbol{x}}^n[f](x) - C_{\boldsymbol{x}}[f](x) = -\sum_{|k|>n} f(kh)S_h(x_k, x), \\ \check{E}T_{\boldsymbol{x}}^n[f] &:= T_{\boldsymbol{x}}^n[f] - T_{\boldsymbol{x}}[f] \qquad = -h\sum_{|k|>n} f(kh), \end{aligned}$$

and call them *the truncation errors*. The truncation errors are the sum of truncated terms for $|k| > n$ and dominated by the half width R of the approximate support.

The errors of the finite Sinc approximations

$$\begin{aligned} Ec_x^n[f](x) &:= c_x^n[f](x) - f(x) &= Ec_x[f](x) + \check{E}c_x^n[f](x), \\ EC_x^n[f](x) &:= C_x^n[f](x) - F(x) &= EC_x[f](x) + \check{E}C_x^n[f](x), \\ ET_x^n[f] &:= T_x^n[f] - I[f] &= ET_x[f] + \check{E}T_x^n[f], \end{aligned}$$

are the sum of the discretization errors and the truncation errors.

We define the uniform norm of a function f on the real line as

$$\|f\|_\infty = \sup_{x \in \mathbb{R}} |f(x)| .$$

The errors of the finite Sinc approximations are bounded by the triangle inequalities

$$\begin{aligned} \left\|Ec_x^n[f]\right\|_\infty &\le \|Ec_x[f]\|_\infty + \left\|\check{E}c_x^n[f]\right\|_\infty, \\ \left\|EC_x^n[f]\right\|_\infty &\le \|EC_x[f]\|_\infty + \left\|\check{E}C_x^n[f]\right\|_\infty, \\ \left|ET_x^n[f]\right| &\le |ET_x[f]| + \left|\check{E}T_x^n[f]\right| . \end{aligned} \tag{7.17}$$

The discretization errors and the truncation errors are bounded by the following theorem.

Theorem 7.10 (Shiraishi and Sugiura 2018) *If $f \in \mathbf{G}(A, \alpha, \beta)$, the discretization errors are bounded by*

$$\begin{aligned} \|Ec_x[f]\|_\infty &\le \frac{4A\beta}{\pi\sqrt{\pi\alpha}(1 - e^{-\pi^2/(\beta h^2)})} h e^{-\pi^2/(4\beta h^2)}, \\ \|EC_x[f]\|_\infty &\le \frac{2A\beta}{\sqrt{\pi\alpha}(1 - e^{-\pi^2/(\beta h^2)})} h^2 e^{-\pi^2/(4\beta h^2)}, \\ |ET_x[f]| &\le \frac{2\sqrt{\pi} A}{\sqrt{\alpha}(1 - e^{-2\pi^2/(\beta h^2)})} e^{-\pi^2/(\beta h^2)}. \end{aligned} \tag{7.18}$$

And the truncated errors are bounded by

$$\begin{aligned} \left\|\check{E}c_x^n[f]\right\|_\infty &\le \frac{2A}{1 - e^{-\alpha Rh}} e^{-\alpha R^2}, \\ \left\|\check{E}C_x^n[f]\right\|_\infty &\le \frac{8A}{\pi(1 - e^{-\alpha Rh})} h e^{-\alpha R^2}, \\ \left|\check{E}T_x^n[f]\right| &\le \frac{2A}{1 - e^{-\alpha Rh}} h e^{-\alpha R^2}. \end{aligned} \tag{7.19}$$

The discretization errors in (7.18) rapidly decrease to zero in the order $O(r^{h^{-2}})$ as $h \to 0$, where $r = e^{-\pi^2/(4\beta)}$ for the Sinc interpolation and for the Sinc integral, and $r = e^{-\pi^2/\beta}$ for the trapezoidal rule. The truncation error in (7.19) rapidly decrease to zero in the order $O(r^{R^2})$, $r = e^{-\alpha}$, as $R \to \infty$. The Gauss parameters α and β dominate the truncation errors and the discretization errors, respectively.

7.1.5 DE Formula (Double Exponential Formula)

We consider approximating the integral of an analytic function $f(x)$ on a semi-infinite interval $[a, \infty)$,

$$I_S[f] := \int_a^{\infty} f(x)dx.$$

Because $f(x)$ is not a rapidly decreasing function on $(-\infty, \infty)$, we cannot use the trapezoidal rule directly. So, we transform the integral to an integral on the infinite interval $(-\infty, \infty)$ with the variable transformation $x = \psi(y)$, $\psi(-\infty) = a$, $\psi(\infty) = \infty$,

$$I[f_1] = \int_{-\infty}^{\infty} f_1(y)dy = I_S[f], \quad f_1(y) = f(\psi(y))\psi'(y),$$

and use the finite trapezoidal rule. The resulted integral formula is

$$T_{\boldsymbol{x}}^n[f_1] = h \sum_{k=-n}^{n} f_1(x_k) = h \sum_{k=-n}^{n} f(\psi(x_k))\psi'(x_k) \cong I[f_1] = I_S[f],$$

with the equidistributed sample points

$$\boldsymbol{x} = (x_k), \quad x_k = x_0 + kh \ (k \in \mathbb{Z}),$$

where x_0 is the sample origin and $h > 0$ is the sample spacing of the sample points.

In DE formula (Double Exponential formula) presented by Takahashi and Mori (1974), the transformation $x = \psi(y)$ is selected so that $f_1(y)$ decreases double exponentially as $|y| \to \infty$, that means, there exist positive constants $A, \ B > 0$ satisfying

$$|f_1(y)| = O\left(\exp\left(-Ae^{B|y|}\right)\right) \qquad (|y| \to \infty).$$

We call such a variable transformation *DE transformation (Double Exponential transformation)*. Under appropriate conditions, the error of the DE formula is estimated as

$$\left|T_{\boldsymbol{x}}^n[f_1] - I_S[f]\right| = O\left(e^{-CN/\log N}\right) \qquad (|N| \to \infty),$$

where $C > 0$ is a positive constant and $N = 2n + 1$ is the number of sample points. The error of the DE formula is a highly accurate integral formula whose error decreases exponentially with respect to $N/\log N$.

To use DE formula, we must select an appropriate variable transformation $x = \psi(y)$ depending on the integrand $f(x)$. Semi-indefinite integrations in this chapter are

$$I_S[f] = \int_0^\infty U(tx)g(x|m)dx,$$

where $U(x)$ is a distribution function which can be connected analytically to an entire function and $g(m|x)$ is the density function defined in (1.7). When $m > 0$ is big, the integrand $f(x) = U(tx)g(m|x)$ has a sharp peak in the neighborhood of the point $x = 1$ derived from $g(m|x)$, because $g(x|m)$ asymptotically approaches the density function of the normal distribution $\sqrt{m/\pi}\, e^{-m(x-1)^2}$ as $m \to \infty$.

For integrate this peak accurately, we adopt the DE transformation

$$\begin{aligned} \psi(y) &= e^{y+1-e^{-y}} = \exp(y - \exp_1(-y)), \\ \psi'(y) &= (1 + e^{-y})\exp(y - \exp_1(-y)) \end{aligned} \tag{7.20}$$

satisfying $\psi(-\infty) = 0,\ \psi(0) = 1,\ \psi(\infty) = \infty$. Hear, $\exp_1(x) = e^x - 1$. The peak moves into the neighborhood of $y = 0$ and can be handled precisely in the computation.

When x is very closed to zero, the subtraction $e^x - 1$ causes a serious loss of significant digits. On C compilers implemented c99 standard, we can use the common mathematical function `expm1(x)` $=\exp_1(x)$ which enables us to calculate $\exp_1(x)$ without this cancellation of significant digits in the neighborhood of $x = 0$. We recommend to use `expm1(x)` instead of the function `exp(x) − 1` in the calculation of $\psi(y)$ and $\psi'(y)$

The integrand $f(x)$ satisfies

$$\begin{aligned} f(x) &= O\left(e^{-mx}\right) \quad (x \to \infty), \\ f(x) &= O\left(x^{m-1}\right) \quad (x \to 0), \end{aligned}$$

then it holds that

$$\begin{aligned} f_1(y) &= O\left(\exp(-me^{y})\right) \quad (y \to \infty), \\ f_1(y) &= O\left(\exp(-me^{-y})\right) \quad (y \to -\infty), \end{aligned}$$

for every positive number $\delta > 0$. Therefore, $T_x^n[f_1]$ becomes a DE formula.

7.2 Computation of Statistic Values of Hayter Type

In this section, we describe the computation of statistic values of Hayter type used in the multiple comparison procedures in Chaps. 3 and 4.

In Sect. 7.2.1, we explain the nature of the distribution functions $D_1(t|k)$, $TD_1(t|k,m)$.

In Sect. 7.2.2, we describe the algorithms to calculate $D_1(t|k)$, $TD_1(t|k,m)$, and $td_1(k,m;\alpha)$. The algorithm for $D_1(t|k)$ with finite Sinc method is introduced in Sect. 7.2.2.1. The algorithm for $TD_1(t|k,m)$ with DE formula is introduced in Sect. 7.2.2.2. The algorithm for $td_1(k,m;\alpha)$ with the secant method is introduced in Sect. 7.2.2.3. In Sect. 7.2.2.4, we show some numerical examples.

For the calculation of $TD_1(t|k,m)$, we need the function $g(x|m)$ defined by (1.7). The calculation of $g(x|m)$ is difficult when $m \gg 1$, because it has a very sharp peak in the neighborhood of $x=1$ and very rapidly decay in the neighborhood of $x=0,\infty$. We describe how to calculate the $g(x|m)$ in high accuracy in Sect. 7.2.3.

7.2.1 Distribution Functions of Hayter Statistic and Their Nature

The Hayter type distributions defined in Chap. 3 are

$$\begin{aligned} D_1(t|\,k) &:= P\left(\max_{1\le i<i'\le k}\frac{Z_{i'}-Z_i}{\sqrt{2}}\le t\right),\\ TD_1(t|\,k,m) &:= P\left(\max_{1\le i<i'\le k}\frac{Z_{i'}-Z_i}{\sqrt{2U_E/m}}\le t\right), \end{aligned} \tag{7.21}$$

with an integer $k\ge 2$, where $Z_i \sim N(0,1)$ $(1\le i\le k)$ are independent and U_E be a random variable distributed to χ^2-distribution with m degrees of freedom that is independent of $Z_i,\ldots.Z_k$. For $t\ge 0$, by using the recurrence relations

$$\begin{aligned} H_1(t,x) &= \Phi\left(\sqrt{2}\,t+x\right),\\ H_r(t,x) &= \int_{-\infty}^{x} H_{r-1}(t,y)\varphi(y)dy \\ &\quad + H_{r-1}(t,x)\left\{\Phi\left(\sqrt{2}\,t+x\right)-\Phi(x)\right\} \quad (2\le r\le k-1), \end{aligned} \tag{7.22}$$

$D_1(t|\,k)$ is expressed by the infinite integral

$$D_1(t|k) = \int_{-\infty}^{\infty} H_{k-1}(t,x)\varphi(x)dx. \tag{7.23}$$

The relations above enable us to expand the domain of $D_1(t|k)$ to $t \in \mathbb{R}$.

For $t \geq 0,\ 1 \leq m \leq \infty$, the distribution function $TD_1(t|\ k, m)$ is expressed by

$$TD_1(t|k, m) = \int_0^\infty D_1(ts|k)g(s|m)ds \quad (m < \infty), \tag{7.24}$$

$$TD_1(t|k, \infty) = D_1(t|k), \tag{7.25}$$

where $g(s|m)$ is a density function for $\sqrt{U_E/m}$ defined by (1.7). For a real number $0 < \alpha < 1$, the upper $100\alpha\%$ point t^* of the distribution function $TD_1(t|\ k, m)$ denoted by

$$t^* = td_1(k, m; \alpha) \tag{7.26}$$

is a solution of the following equation for t

$$TD_1(t^*|\ k, m) = 1 - \alpha. \tag{7.27}$$

Now, we give the nature of functions relevant to the computations above.

Theorem 7.11 *For the integrand $H_r(t, x)\varphi(x)$ $(r \geq 1)$ in (7.22), (7.23), it holds that*

$$H_r(t,\ \cdot\ )\varphi(\ \cdot\ ) \in \mathbf{G}\left(\frac{1}{\sqrt{2\pi}}, \frac{1}{2}, \frac{r+1}{2}\right) \quad (t \geq 0). \tag{7.28}$$

Theorem 7.12 *For $D_1'(t|k) = dD_1(t|k)/dt$ with $k \geq 2$, it holds that*

$$D_1'(\cdot|k) \in \mathbf{G}\left(\frac{1}{2}, k - 1\right). \tag{7.29}$$

7.2.2 Computation of the Distribution Functions and the Upper 100α % Point*

We give detailed computation methods of the upper $100\alpha^*\%$ point for the multiple comparison methods for average differences explained in Chap. 3.

We make use of finite Sinc integral in (7.22), the trapezoidal rule in (7.23), and DE formula in (7.24). Let $\varepsilon = 10^{-8}$ be a given required accuracy. Then, since the integrand $H_r(t, x)\varphi(x)$ $(1 \leq r \leq k - 1)$ in (7.22), (7.23) is a function quickly decreasing as $|x| \to \infty$ and belongs to the function family $\mathbf{G}$, the finite integrals on the integration intervals such that the function values in magnitude are greater than ε contribute to the integrals (7.22), (7.23). Since $0 \leq H_r(t, x) \leq 1$ for the probability $H_r(t, x)$ and $\varphi(6) = 6.07 \cdots \times 10^{-9}$, for $|x| > 6$, we have

$$|H_r(t, x)\varphi(x)| < \varepsilon \quad (1 \leq r \leq k - 1).$$

So, it suffices to consider the integration interval $I_6 = [-6,\ 6]$. Numerical experiments reveal that for $2 \leq k \leq 10$ the Sinc approximations with the number of sample points $N = 25$ ($n = 12$) and the sample spacing $h = 6/12 = 0.5$ give approximations of the accuracy of around $\varepsilon = 10^{-8}$. To verify the accuracy obtained, we compare the results with those obtained by Sinc approximations with the required error criterion $\varepsilon = 10^{-15}$ smaller than 10^{-8} and the integration interval $I_8 = [-8,\ 8]$ larger than $[-6, 6]$.

7.2.2.1 Computation of $D_1(t|k)$

Define a vector $\boldsymbol{x}$ of sample points on the interval I_6 by

$$\boldsymbol{x} = (x_{-n}, x_{1-n}, \ldots, x_n)^T \in \mathbb{R}^N, \qquad x_j = jh\ (-n \leq j \leq n), \qquad h = 6/n.$$

Denote a vector $\boldsymbol{f}$ of function values $f(x)$ at the sample points above by

$$\boldsymbol{f} = f(\boldsymbol{x}) = (f(x_{-n}),\ f(x_{1-n}), \ldots, f(x_n))^T \in \mathbb{R}^N.$$

For two given sample vectors $\boldsymbol{f}_1$ and $\boldsymbol{f}_2$, define a vector $\boldsymbol{f}_1\boldsymbol{f}_2$ of element-wise products by

$$\boldsymbol{f}_1\boldsymbol{f}_2 = (f_1(x_{-n})f_2(x_{-n}),\ f_1(x_{1-n})f_2(x_{1-n}), \ldots,\ f_1(x_n)f_2(x_n))^T \in \mathbb{R}^N,$$

a vector of the function values $f_1(x)f_2(x)$ at the sample points. Denote the finite Sinc integral composed of the sample vector $\boldsymbol{f}$ by

$$C_{\boldsymbol{x}}^n[\boldsymbol{f}](x) = \sum_{j=-n}^{n} f(x_j)S_h(x_j, x).$$

Then, its value at the sample point x_i is given by

$$C_{\boldsymbol{x}}^n[\boldsymbol{f}](x_i) = \sum_{j=-n}^{n} f(x_j)S_h(x_j, x_i), \quad (-n \leq i \leq n),$$

or in vector notation

$$\begin{aligned} C_{\boldsymbol{x}}^n[\boldsymbol{f}] &= Q\boldsymbol{f}, \qquad Q = (q_{ij}) \in \mathbb{R}^{N\times N}, \\ q_{ij} &= S_h(x_j, x_i) \quad (-n \leq i \leq n,\ -n \leq j \leq n). \end{aligned} \tag{7.30}$$

The linear operator defined by the matrix Q is a discretization of the indefinite integral operator.

From the discussion above, we obtain an algorithm in a concise form for computing the vector $\boldsymbol{H}_{k-1}$ of function values for $H_{k-1}(t, x)$ and $D_1(t|k)$ as follows.

<Algorithm: function values $D_1(t|k)$ >

(1) Initialization (before inputting t and k): store the matrix Q given by (7.30) in an array.

(2) Initialization (after imputing t and k):

  (1) compute the vectors $\boldsymbol{\varphi}$, $\boldsymbol{\Phi}_s$, and $\boldsymbol{\Phi}_\Delta$ of $\varphi(x)$, $\Phi(\sqrt{2}\,t+x)$, and $\Phi(\sqrt{2}\,t+x)-\Phi(x)$ and store them in arrays.

  (2) Store the vector $\boldsymbol{H}_1 = \boldsymbol{\Phi}_s$ of $H_1(t,x)$ in an array.

(3) Recursion: computation of the vector $\boldsymbol{H}_r$ of $H_r(t,x)$ for $r = 2, 3, \ldots, k-1$.

  (1) Compute the vector $\boldsymbol{u}_r = \boldsymbol{H}_{r-1}\boldsymbol{\varphi}$ of $u_r(x) = H_{r-1}(t,x)\varphi(x)$.

  (2) For the indefinite integral $U_r(x)$ of $u_r(x)$, compute the vector $\boldsymbol{U}_r = Q\boldsymbol{u}_r$.

  (3) Compute the vector $\boldsymbol{H}_r = \boldsymbol{U}_r + \boldsymbol{H}_{r-1}\boldsymbol{\Phi}_\Delta$ of $H_r(x)$ and store it in an array.

(4) Computation of $D_1(t|k)$ by the finite trapezoidal rule.

  (1) Compute the vector $\boldsymbol{u}_k = \boldsymbol{H}_{k-1}\boldsymbol{\varphi}$ of $u_k(x) = H_{k-1}(t,x)\varphi(x)$.

  (2) Compute $D_1(t|k) = h\sum_{i=-n}^{n}(\boldsymbol{u}_k)_i$ and output it.

  Recall that $\boldsymbol{H}_{r-1}\boldsymbol{\varphi}$ and $\boldsymbol{H}_{r-1}\boldsymbol{\Phi}_\Delta$ in (3) and (4) mean the element-wise products of the vectors. Note that $(\boldsymbol{u}_k)_i$ in (4) is the ith element of $\boldsymbol{u}_k$ and an approximation to $u_k(x_i) = H_{k-1}(t,x_i)\varphi(x_i)$.

This algorithm enables us to construct a function program `D1[t, k]` in Mathematica for computing $D_1(t|k)$ with input parameters `t` $= t$ and `k` $= k$ and the output $D_1(t|k)$.

7.2.2.2 Computation of $TD_1(t|k,m)$

In this subsection, using the function program `D1[t, k]` given above, we compute $TD_1(t|k,m)$ with the DE formula in Sect. 7.1.5. In view of the law of large numbers, the factor $g(s|m)$ in the integrand in (7.24) approaches to the normal distribution $\sqrt{m}\varphi\left(\sqrt{m}(s-1)\right)$ as $m \to \infty$. The integrand in itself approaches to $\sqrt{m}D_1(t|k)\varphi\left(\sqrt{m}(s-1)\right)$ as m grows. Therefore, we choose the spacing $h = 0.15/\sqrt{m}$ of the sample points for the DE formula in Sect. 7.1.5 so that we approximate $\sqrt{m}\varphi\left(\sqrt{m}(s-1)\right)$ to a sufficient accuracy.

This procedure allows us to construct a function program `TD1[t, k, m]` in Mathematica for computing $TD_1(t|k,m)$ for $t \ge 0$, $0 \le k \le 10$ and $1 \le m \le \infty$ with input parameters `t` $= t$, `k` $= k$, and `m` $= m$ and the output $TD_1(t|k,m)$. Mathematica allows the input $m = \infty$ by `m = Infinity` to obtain the output $TD_1(t|k,\infty) = D_1(t|k)$ in (7.25).

7.2.2.3 Computation of $td_1(k,m;\alpha)$, the Upper 100α% Point

For $k \ge 2$, $1 \le m \le \infty$ and $0 < \alpha < 1$, the upper 100α% point $t^\star = td_1(k, m;\alpha)$ is a solution of the nonlinear equation for t,

$$f(t) = TD_1(t|k, m) = 1 - \alpha.$$

The well-known Newton method for the solution of nonlinear equations is not available here, since the derivative of $f(t)$ is required. Instead, we make use of the secant method, where with two given starting values $t_0,\ t_1 \cong t^\star$, we construct a sequence $\{t_k\}_{k\geq 0}$ converging to the solution $t^\star$ by the recurrence relation

$$t_{k+1} = t_k - \Delta t_k, \quad \Delta t_k = \frac{f(t_k)(t_k - t_{k-1})}{f(t_k) - f(t_{k-1})}, \qquad k = 1,\ 2, \ldots. \tag{7.31}$$

It is well known that the order of convergence of the secant method is the golden ratio $\phi = (1 + \sqrt{5})/2 = 1.61\cdots$ (Ralston and Rabinowitz 1978), namely, for the sequence $\{t_k\}$ converging to the solution $t^\star$, we have

$$\left|t_{k+1} - t^\star\right| = O\left(\left|t_k - t^\star\right|^\phi\right) \quad (k \to \infty).$$

Although the order of convergence ϕ of the secant method is smaller than the order two of the Newton method, the secant method requires only one function evaluation per iteration that is smaller in number than the Newton method with evaluation of the function and its derivative.

We constructed a function program `td1[k,m,alp,t1]` in Mathematica for computing $td_1(k, m; \alpha)$ using the secant method with the function program `TD1[t,k,m]` given above. The input parameters are `k` $= k$, `m` $= m$, `alp` $= \alpha$, and `t1` $= t_1$, where t_1 is one of the starting values of secant method and another one $t_0 = t_1 - 10^{-3}$ is generated automatically in the program. If $|\Delta t_k| < 10^{-12}$, then we stop the iteration of the secant method and obtain the solution t_{k+1}.

It is advisable to choose the starting value t_1 smaller than the solution $t^\star$ by the reason shown below. Near the solution, $TD_1(t|k, m)$ is a concave and monotonically increasing function and converges quickly to 1 as $t \to \infty$. So, if $t_1 < t^\star$, then t_k monotonically approaches to the solution $t^\star$ on the real line. Otherwise, if t_1 is much larger than the solution $t^\star$, then for $k = 1$, $f(t_1) - f(t_0) \cong 1 - 1 = +0$, the denominator of Δt_1 in (7.31), so $\Delta t_1 \cong +\infty$ and t_2 goes to $-\infty$.

7.2.2.4 Results of Numerical Computation

The numerical computations in this book are performed with Mathematica ver. 8.0.4 on the computer Apple iMac(2.8 GHz, Intel Core i7), OSX10.9.5.

We computed the values in Table 6.4 in Sect. 6.4. We use the function program `td1[k, m, alp, t1]` shown above to compute 37 upper $100\alpha^\star$% points.

In this computation, the function program `td1[k, m, alp, t1]` is called 37 times with 14.2 s, the average computation time being 0.38 s, and maximum error in magnitude 5.5×10^{-9} with six iterations in the secant method in average. In total, the function program `TD1[t, k, m]` is called 295 times with 4.8×10^{-2} s in average computation time.

To verify the error above, we performed the computation again with higher required accuracy, namely, the error criterion for the Sinc approximation being $\varepsilon = 10^{-15}$, the support $I_8 = [-8, 8]$, the number of sample points $N = 49$ $(n = 24)$, and the spacing $h = 0.1/\sqrt{m}$ of the sample points for the DE formula in Sect. 7.1.5 with 45.4 s in total computation time. The error shown above is estimated by the difference between this higher accuracy result and the above one.

The use of some compiler language such as Fortran or C could perform the numerical computation at higher speed than the use of Mathematica that is an interpreter language.

7.2.3 Numerical Computation of the Density Function $g(s|m)$

We describe the computation of the density function (1.7)

$$g(x|m) = \frac{2(m/2)^{m/2}}{\Gamma(m/2)} x^{m-1} e^{-mx^2/2} \quad (x \geq 0),$$

with C programming language (C99).

Since this function asymptotically approaches to the density function of the normal distribution $\sqrt{m/\pi}\ e^{-m(x-1)^2}$ as $m \to \infty$, this function has a single sharp peak near $x = 1$ for large m. Change of variables so that the point $x = 1$ shifts to the origin enables us to compute this peak in high accuracy, as shown below.

The floating-point numbers discretely distribute on the real line. The distance of two neighboring floating-point numbers is approximately equal to 2×10^{-16} near $x = 1$ and 5×10^{-324} near the origin.

In practice, we define a function $g_1(x|m)$ by

$$\begin{aligned} g_1(x|m) &= g(1 + x|m) \\ &= \frac{2(m/2)^{m/2}}{\Gamma(m/2)} (1 + x)^{m-1} e^{-m(1+x)^2/2} \quad (x \geq -1). \end{aligned} \tag{7.32}$$

For large m, the computations of the factors $\Gamma(m/2)$, $(m/2)^{m/2}$, $(1 + x)^{m-1}$, and $\exp(-m(1 + x)^2/2)$ in the right of (7.32) suffer from overflow or underflow and thus the computed results have low accuracies or no significant digits. To avoid this difficulty, we adopt the following procedure:

$$g_1(x|m) = \sqrt{\frac{m}{\pi}} e^X,$$

$$X = \log g_1(x|m) - \frac{1}{2} \log(m/\pi) = Y + Z,$$

$$Y = -\frac{1}{2}\left\{2\log\Gamma\left(\frac{m}{2}\right) - m\log\frac{m}{2} + m + \log\frac{m}{4\pi}\right\}, \tag{7.33}$$

$$Z = -\frac{m}{2}x(2+x) + (m-1)\log(1+x). \tag{7.34}$$

To ensure high accuracy in the computed results, we perform the computation with the extended double precision (the unit roundoff $\mathrm{u}_E = 2^{-60}$), higher in accuracy by seven bits than the double precision (the unit roundoff $\mathrm{u} = 2^{-53}$). We use the long double function `lgammal(x)` $= \log\Gamma(x)$ in C99 to compute $\log\Gamma(x)$. For computing $\log(1+x)$ we use `log1pl(x)` $= \log(1+x)$ in order to ensure high relative accuracy of $\log(1+x)$ near the origin $x = 0$.

The loss of significant digits in the computations of Y and Z happens in this procedure for large m. Although the problem of overflow or underflow is avoided by using the logarithmic function, the difficulty occurs in another type, loss of significant digits.

We now give a brief explanation of *loss of significant digits*. For given numbers a and b, in the addition or subtraction, $c = a \pm b$, the number of significant digits (the number of correct digits) for c is smaller than the number of digits used in the computation when $|c|$ is smaller than the values of $|a|$ and $|b|$. We call this phenomenon *loss of significant digits*. For a bigger value of $|a|/|c|$, more significant digits are lost. Specifically, loss of significant digits happens when $a \cong b$ in subtraction and $a \cong -b$ in addition.

We give an example. For $a = 1.2345678\ldots$ and $b = 1.2343210\cdots$, we obtain $c = a - b = 0.000246\ldots$. We now perform this computation in four significant digits, namely, a and b are rounded to values with four digits. Then, we obtain an approximation $c' = 1.235 - 1.234 = 0.001$ with no significant digits. The computation with six digits gives $c' = 1.23457 - 1.23432 = 0.0025$ with only two significant digits. The number of significant digits lost in this computation is $4 \cong \log_{10}|a|/|c|$, the number of digits coincident of a and b.

1. Strategy for avoiding loss of significant digits occurring in the computation of Y with (7.33) to obtain the value of e^Y in the accuracy of double precision.

By numerical experiments, we verified that although for $1 \le m \le 42$, loss of significant digits happens in the right-hand side of (7.33) in the extended double-precision computation, the error of e^Y is within the unit roundoff u of double precision.

For a large value of $m > 42$, we make use of a truncated approximation of the asymptotic expansion of $\log\Gamma(x)$ as follows:

$$G_n(x) := (x - \tfrac{1}{2})\log x - x + \tfrac{1}{2}\log(2\pi) + \sum_{k=1}^{n}\frac{B_{2k}\,x^{-2k+1}}{2k(2k-1)} \tag{7.35}$$
$$\cong \log\Gamma(x) \quad (x > 0),$$

where B_k is the Bernoulli number. The error of $G_n(x)$ is bounded by

$$|G_n(x) - \log \Gamma(x)| \leq \frac{|B_{2n+2}\,x^{-2n-1}|}{(2n+1)(2n+2)} \quad (x > 0). \tag{7.36}$$

For large n and $x > 0$, we obtain an accurate approximation to $\log \Gamma(x)$ (cf. Abramowitz and Stegun p. 257 **6.1.42** (1972)). Substituting (7.35) into (7.33), we obtain an approximation with no loss of significant digits to Y as follows:

$$\begin{aligned} Y_n &= -\frac{1}{2}\left\{2G_n\left(\frac{m}{2}\right) - m\log\frac{m}{2} + m + \log\frac{m}{4\pi}\right\} \\ &= -\sum_{k=1}^{n} \frac{B_{2k}}{2k(2k-1)}\left(\frac{2}{m}\right)^{2k-1} \cong Y. \end{aligned}$$

In view of (7.36), the error is bounded by

$$\begin{aligned} |Y_n - Y| &= \left|\log \Gamma\left(\frac{m}{2}\right) - G_n\left(\frac{m}{2}\right)\right| \\ &\leq \frac{|B_{2n+2}|}{(2n+1)(2n+2)}\left(\frac{2}{m}\right)^{2n+1} \\ &= O(m^{-2n-1}) \quad (m \to \infty). \end{aligned} \tag{7.37}$$

Since $|Y_n - Y| \leq 5.5 \times 10^{-18} \leq \mathrm{u}$ for $m > 42$ and $n = 5$, we obtain an approximation Y_n in the accuracy of double precision.

Finally, we obtain an approximation to Y as follows:

$$Y \cong \begin{cases} -\frac{1}{2}\left\{2\log\Gamma\left(\frac{m}{2}\right) - m\log\frac{m}{2} + m + \log\frac{m}{4\pi}\right\} & (1 \leq m \leq 42), \\ Y_5 = -\sum_{k=1}^{5} \frac{B_{2k}}{2k(2k-1)}\left(\frac{2}{m}\right)^{2k-1} & (m > 42), \end{cases}$$

where the Bernoulli numbers are given by

$$B_2 = \frac{1}{6},\ B_4 = -\frac{1}{30},\ B_6 = \frac{1}{42},\ B_8 = -\frac{1}{30},\ B_{10} = \frac{5}{66}.$$

All computations here are performed in the extended double precision. We use the long double function `lgammal(x)` $= \log \Gamma(x)$ in C99 to compute $\log \Gamma(x)$.

Note that since in view of (7.37), $Y_5 \to 0$ $(m \to \infty)$, we obtain

$$Y \to 0 \ (m \to \infty). \tag{7.38}$$

2. Strategy for avoiding loss of significant digits in the computation of Z with (7.34) to obtain an approximation to e^Z in the accuracy of double precision.

We start by analyzing the behavior of the function Z of x defined by (7.34). For $m \geq 2$, we list the behavior of the function Z of x as follows:

x	-1	$\cdots$	$x_{\max}$	$\cdots$	∞
Z	$-\infty$	$\nearrow$	$Z_{\max}$	$\searrow$	$-\infty$

where $x_{\max}$ and $Z_{\max}$ are defined by

$$x_{\max} = -\frac{1}{m + \sqrt{m(m-1)}} \cong -\frac{1}{2m} \qquad (m \geq 2),$$
$$Z_{\max} = \frac{1}{2}\left\{1 + (m-1)\log\left(1 - \frac{1}{m}\right)\right\} \to 0, \qquad (m \to \infty),$$

respectively. We see that Z is a function having a peak at $x = x_{\max}$.

By change of variables $x = m^{-1/2}t$, we have Maclaurin series of Z with respect to the variable t,

$$Z = -\left(1 - \frac{1}{2}m^{-1}\right)t^2 - m^{-1/2}t - (m-1)\sum_{k=3}^{\infty}\frac{(-1)^k}{k}m^{-k/2}t^k.$$

Let a be a real number such that $0 < a < 1$. Then, it follows that for $t \in [-a, a]$

$$Z = -t^2 + O(m^{-1/2}) \to -t^2 \quad (m \to \infty).$$

From this relation and in view of (7.38), for $t \in [-a, a]$ we have

$$\sqrt{\frac{\pi}{m}}g_1(m^{-1/2}t|m) = e^{Y+Z} \to e^{-t^2} \quad (m \to \infty).$$

This implies that $g_1(x|m)$ asymptotically approaches to the density function of the normal distribution with mean zero and variance $1/(2m)$.

Now, let us consider the computation of the right-hand side of (7.34) for Z. No loss of significant digits happens when $x \to -1 + 0$ and $x \to \infty$, since the dominant term is $\log(1 + x)$ in the former case and $x(2 + x)$ in the latter case. A serious problem occurs in the computation near the origin $x = 0$.

Since for $|x|$ near zero, the first and second terms of the right-hand side of (7.34) and Z are given by

$$\begin{aligned} t_1 &= -\tfrac{1}{2}mx(2+x) &&\cong -mx, \\ t_2 &= (m-1)\log(1+x) &&\cong (m-1)x, \\ Z &= t_1 + t_2 &&\cong -x, \end{aligned}$$

respectively, we have $|t_1|/|Z| \cong m$. Therefore, a serious loss of significant digits happens in the computation of (7.34) for large m.

Computation of the right-hand side of (7.34) in the extended double precision gives e^Z with the absolute error $\leq$ u for $1 \leq m \leq 22000$. For $m > 22000$ and $|x| >$

$8.6/\sqrt{2m}$, the computation of (7.34) in the extended double precision is effective as well. We verify these facts by numerical computation.

For $m > 22000$ and $|x| \leq 8.6/\sqrt{2m}$, by the Maclaurin series of $\log(1+x)$ we have

$$Z = -\frac{2m-1}{2}x^2 - x + (m-1)L,$$
$$L = \log(1+x) - x + \frac{x^2}{2} = \sum_{k=3}^{\infty} \frac{(-1)^{k+1}}{k} x^k.$$

We approximate L by a truncation L_n of the Maclaurin series above for L,

$$L_n = \sum_{k=3}^{n} \frac{(-1)^{k+1}}{k} x^k \cong L.$$

We have a sufficiently accurate approximation to Z using L_n with $n = 10$, for $m > 22000$, so $|x| \leq 8.6/\sqrt{2m} \leq 8.6/\sqrt{2 \times 22000} < 0.041$.

Now, we summarize the computational procedure for obtain Z in the accuracy of double precision.

- For $1 \leq m \leq 22000$ or $|x| > 8.6\sqrt{2m}$,

$$Z \cong -\frac{m}{2}x(2+x) + (m-1)\log(1+x),$$

- for $m > 22000$ and $|x| \leq 8.6\sqrt{2m}$,

$$Z \cong -\frac{2m-1}{2}x^2 - x + (m-1)\sum_{k=3}^{10} \frac{(-1)^{k+1}}{k} x^k.$$

Here, we perform the extended double precision computation and use the function `log1pl(x)` $= \log(1+x)$ in C99 to compute $\log(1+x)$.

7.3 Computation of Level Probability

We explain the computation of *the level probability* $P(L, k; \boldsymbol{\lambda}_n)$ used in Chaps. 3 and 4. We start by extending the domain for the vector variable $\boldsymbol{\lambda}_n$ in the definition in Sect. 3.1 to the real vector from the rational number vector as follows:

$$P(L, k; \boldsymbol{w}) \quad (1 \leq L \leq k, \quad \boldsymbol{w} = (w_1, w_2, \ldots, w_k) \in \mathbb{R}_+^k),$$

where $\mathbb{R}_+ = (0, \infty)$. We say $\boldsymbol{w}$ *a weight vector*. Now that the domain is extended, we here verify the definition $P(L, k; \boldsymbol{w})$.

Let $X_i \sim N(0, 1/w_i)$ $(1 \le i \le k)$ be k independent random variables. Denote by $(\tilde{\mu}_1^*, \tilde{\mu}_2^*, \ldots, \tilde{\mu}_k^*)$ the solution $(\mu_1, \mu_2, \ldots, \mu_k) \in \mathbb{R}^k$ that minimizes the target function

$$f(\mu_1, \mu_2, \ldots, \mu_k) = \sum_{i=1}^{k} w_i (\mu_i - \bar{X}_i)^2,$$

under the condition

$$\mu_1 \le \mu_2 \le \cdots \le \mu_k.$$

In other words, we have

$$\sum_{i=1}^{k} w_i (\tilde{\mu}_i^* - \bar{X}_i)^2 = \min_{\mu_1 \le \mu_2 \le \cdots \le \mu_k} \left\{ \sum_{i=1}^{k} w_i (\mu_i - \bar{X}_i)^2 \right\}.$$

Then, $P(L, k; \boldsymbol{w})$ gives the probability when $\tilde{\mu}_1^* \le \tilde{\mu}_2^* \le \cdots \le \tilde{\mu}_k^*$ take L values that are different from each other.

From this definition, we see that the level probability is invariant for a real multiplication of the weight vector, namely, for an arbitrary $a > 0$,

$$P(L, k; a\boldsymbol{w}) = P(L, k; \boldsymbol{w}).$$

In Chaps. 3 and 4, for a given weight vector $\boldsymbol{w} = (w_1, w_2, \ldots, w_k)$, we need such a level probability

$$P(L, m; \boldsymbol{w}') \ \ (1 \le L \le m),$$

which has a partial weight vector of arbitrary consecutive elements,

$$\boldsymbol{w}' = (w_s, w_{s+1}, \ldots, w_{s+m-1}), \ 1 \le s \le k, \ 1 \le m \le k - s + 1.$$

In this section, we show an algorithm for constructing the table of the level probability

$$P[L, m, s] = P\left(L, m; (w_s, w_{s+1}, \ldots, w_{s+m-1})\right), \qquad (7.39)$$
$$1 \le L \le m \le k - s + 1, \ 1 \le s \le k,$$

for a given weight vector $\boldsymbol{w}$.

7.3.1 Fundamental Algorithm

If $\boldsymbol{w} = (1, 1, \ldots, 1) \in \mathbb{R}_+^k$, then we compute $P(L, k) := P(L, k; \boldsymbol{w})$ using the recurrence relation

$$
\begin{aligned}
P(1,k) &= \frac{1}{k},\\
P(L,k) &= \frac{1}{k}\{(k-1)P(L,k-1) + P(L-1,k-1)\} \quad (2 \le L \le k-1),\\
P(k,k) &= \frac{1}{k!}
\end{aligned}
\tag{7.40}
$$

(Barlow et al. 1972).

For $1 \le k \le 4$, we compute $P(L,k;\boldsymbol{w})$ by the following formulae (Robertson et al. 1988):

$$
\begin{aligned}
&P(1,1;\boldsymbol{w}) = 1,\\
&P(1,2;\boldsymbol{w}) = \tfrac{1}{2}, \quad P(2,2;\boldsymbol{w}) = \tfrac{1}{2},\\
&P(1,3;\boldsymbol{w}) = \tfrac{1}{4} + A, \ P(2,3;\boldsymbol{w}) = \tfrac{1}{2}, \quad P(3,3;\boldsymbol{w}) = \tfrac{1}{4} - A,\\
&P(1,4;\boldsymbol{w}) = \tfrac{1}{8} + B, \ P(2,4;\boldsymbol{w}) = \tfrac{3}{8} + C,\\
&P(3,4;\boldsymbol{w}) = \tfrac{3}{8} - B, \ P(4,4;\boldsymbol{w}) = \tfrac{1}{8} - C,
\end{aligned}
\tag{7.41}
$$

where

$$
\begin{aligned}
A &= \frac{\rho(w_1,w_2,w_3)}{2\pi},\\
B &= \frac{\rho(w_1+w_2,w_3,w_4) + \rho(w_1,w_2+w_3,w_4) + \rho(w_1,w_2,w_3+w_4)}{4\pi},\\
C &= \frac{\rho(w_1,w_2,w_3) + \rho(w_2,w_3,w_4)}{4\pi},\\
\rho(a,b,c) &= \arcsin\sqrt{\frac{ac}{(a+b)(b+c)}}.
\end{aligned}
$$

Although in Robertson et al. (1988) there is a formula for $k = 5$, we do not use it owing to the computation of a definite integral, since then a new algorithm and its error analysis are required.

For an arbitrary k, we use the recurrence relation (Robertson et al. 1988). We start by using an auxiliary table

$$
\begin{aligned}
Q[s_0,s_1,\ldots,s_m] &= P\left(m,m;(u_0,u_1,\ldots,u_{m-1})\right), \\
u_i &= \sum_{j=s_i}^{s_{i+1}-1} w_j \quad (1 \le i \le m),\\
1 \le m \le k, &\ 1 \le s_0 < s_1 < \cdots < s_m \le k+1.
\end{aligned}
\tag{7.42}
$$

When $L = k$, from the table Q it immediately follows that

$$
P(k,k;\boldsymbol{w}) = Q\left[1,2,\ldots,k+1\right].
$$

For $2 \leq L \leq k-1$, we use the recurrence relation

$$P(L,k;\boldsymbol{w}) = \sum_{1=s_0<s_1<\cdots<s_L=k+1} Q[s_0, s_1, \ldots, s_L] \prod_{i=0}^{L-1} P(1, m_i, \boldsymbol{w}_i), \tag{7.43}$$
$$m_i = s_{i+1} - s_i,$$
$$\boldsymbol{w}_i = (w_{s_i}, w_{s_i+1}, \ldots, w_{s_{i+1}-1}) \quad (0 \leq i \leq L-1),$$

where the summation is made over the integer sequence $1 = s_0 < s_1 < \cdots < s_L = k+1$ of the length $L+1$.

Finally, for $L = 1$, we obtain $P(1,k;\boldsymbol{w})$ by

$$P(1,k;\boldsymbol{w}) = 1 - \sum_{L=2}^{k} P(L,k;\boldsymbol{w}). \tag{7.44}$$

7.3.2 Computation of the Table Q

Denote the integer sequence $1 \leq s_0 < s_1 < \cdots < s_m \leq k+1$ by a vector

$$\boldsymbol{s}_m = (s_0, s_1, \ldots, s_m).$$

Let $\varphi(\sigma, x)$ be a density function of the normal distribution $N(0, \sigma^2)$. Then, the level probability $Q[\boldsymbol{s}_m]$ is computed by the recurrence relation

$$f_1(x) = \varphi\left(u_0^{-1/2}, x\right),$$
$$f_i(x) = \varphi\left((u_{i-1}^{-1/2}, x\right) \int_{-\infty}^{x} f_{i-1}(t)dt \quad (i = 2, 3, \ldots, m),$$
$$Q[\boldsymbol{s}_m] = \int_{-\infty}^{\infty} f_m(s)ds, \tag{7.45}$$

(Hayter and Liu 1996), where

$$u_i = \sum_{j=s_i}^{s_{i+1}-1} w_j \quad (0 \leq i \leq m-1).$$

Letting $f_m(t) = f_{s_m}(x)$ in (7.45), we obtain the recurrence relation

$$f_{s_1}(x) = \varphi(u_0^{-1/2}, x), \qquad u_0 = \sum_{j=s_0}^{s_1-1} w_j,$$

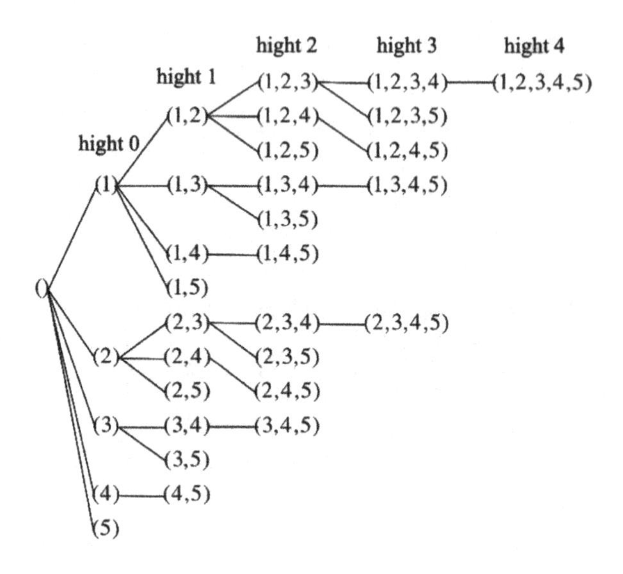

Fig. 7.10 Computation tree for $k = 4$

$$f_{s_m}(x) = \varphi(u_{m-1}^{-1/2}, t) \int_{-\infty}^{x} f_{s_{m-1}}(t)dt, \qquad u_{m-1} = \sum_{j=s_{m-1}}^{s_m - 1} w_j, \tag{7.46}$$

and

$$Q[\mathbf{s}_m] = \int_{-\infty}^{\infty} f_{s_m}(t)dt. \tag{7.47}$$

To compute (7.46) and (7.47), we make use of a computation tree with vertices $\mathbf{s}_m$ and edges $\mathbf{s}_{m-1} \rightarrow \mathbf{s}_m$. We say $m = \dim \mathbf{s}_m - 1$ the height of the vertex $\mathbf{s}_m$. Figure 7.10 shows the computation tree for $k = 4$ with the tree height 4. The height of the root ( ) is -1 under the ground.

On the computation tree, the vertex $\mathbf{s}_m$ computes $f_{s_m}(x)$ of the recurrence relation (7.46) and stores it. Further, the vertex determines the value of $Q[\mathbf{s}_m]$ in (7.47).

This computation requires smaller storage owing to an efficient algorithm of updating the routes on the computation tree. Denote by $p = \langle s_0, s_1, \ldots, s_m \rangle$ the route from the root ( ) to the vertex $\mathbf{s}_m = (s_0, s_1, \ldots, s_m)$,

$$p = \langle ( ) \rightarrow \mathbf{s}_0 \rightarrow \mathbf{s}_1 \rightarrow \cdots \rightarrow \mathbf{s}_m \rangle,$$
$$\mathbf{s}_i = (s_0, s_1, \ldots, s_i) \quad (0 \leq i \leq m).$$

We say the height m of $\mathbf{s}_m$ a route height.

< **Algorithm** $Q[s_0, s_1, \ldots, s_m]$ >

The sequence of functions $\left(f_{\mathbf{s}_i}(x)\right)_{1 \leq i \leq m}$ on the route p is stored in the array $(f_i(x))_{1 \leq i \leq k}$ of length k in the form $f_i(x) = f_{\mathbf{s}_i}(x)$. We actually store the Sinc coefficients of $f_{\mathbf{s}_i}(x)$ or the function values at the sample points, for $f_{\mathbf{s}_i}(x)$.

(0) Let the initial route be $p = \langle s_0 \rangle$ with $s_0 = 1$ and the initial height $m = 0$.
(1) The indefinite integral at the top vertex s_m of $p = \langle s_0, s_1, \ldots, s_m \rangle$.

(1-0) If $m = 0$ or $(k+1) - s_m + m \leq 4$, then go to (2).
(If $(k+1) - s_m + m \leq 4$, the route length is less than or equal to 5)
(1-1) When $m = 1$, compute

$$f_{s_1}(x) = \varphi\left(u_0^{-1/2}, x\right), \quad u_0 = \sum_{j=s_0}^{s_1-1} w_j,$$

and store it in $f_1(x)$ ($f_1(x) = f_{s_1}(x)$).
(1-2) When $m \geq 2$, using finite Sinc integral with $f_{m-1}(x) = f_{s_{m-1}}(x)$, compute

$$f_{s_m}(x) = \varphi\left(u_{m-1}^{-1/2}, t\right) \int_{-\infty}^{x} f_{m-1}(t)dt, \tag{7.48}$$

$$u_{m-1} = \sum_{j=s_{m-1}}^{s_m-1} w_j,$$

and store them in $f_m(x)$ ($f_m(x) = f_{s_{m+1}}(x)$).
(2) The definite integral at the top vertex s_m of $p = \langle s_0, s_1, \ldots, s_m. \rangle$.

(2-0) When $m = 0$, jump to (3).
(2-1) When $1 \leq m \leq 4$, using (7.41), compute

$$Q[s_m] = P(m, m, (u_0, u_1, \ldots, u_{m-1})],$$

$$u_i = \sum_{j=s_i}^{s_{i+1}-1} w_j \quad (0 \leq i \leq m-1).$$

(2-2) When $m \geq 5$, using the finite trapezoidal rule, compute

$$Q[s_m] = \int_{-\infty}^{\infty} f_m(t)dt. \tag{7.49}$$

(3) Updating the route

(3-1) When $s_m < k+1$,
extend the route to $p = \langle s_0, s_1, \ldots, s_m, s_{m+1} \rangle$, $s_{m+1} = s_m + 1$.
(the route length is extended by one.)
(3-2) When $s_m = k+1$,
change the route to $p = \left\langle s_0, s_1, \ldots, s_{m-2}, s'_{m-1} \right\rangle$, $s'_{m-1} = s_{m-1} + 1$.
(the route height is shortened by one.)

(4) Stopping criterion: If $m = 0$ and $p = \langle k+1 \rangle$, then stop the computation. Otherwise, return to (1).

Let us examine the amount of computations required. Remark that the number of vertices with height m in the computation tree (Fig. 7.10) is

$$\binom{k+1}{m+1}.$$

No computation is performed between the root ( ) and the vertex with height 0. The definite integral (7.49) is not required at the vertices with heights $1 \leq m \leq 4$, since the formula (7.41) is used. Therefore, the number of definite integrals required is

$$2^{k+1} - 1 - \sum_{m=0}^{4} \binom{k+1}{m+1} = 2^{k+1} + o(2^k) \quad (k \to \infty).$$

The number of the indefinite integrals (7.48) does not exceed the number of vertices 2^{k+1} in the computation tree. Further, at least the same number of the indefinite integrals as the one of the definite integrals is required. To sum up, the number of the indefinite integrals required is $2^{k+1} + o(2^k)$.

7.3.3 Computation of the Table P

We give a procedure for computing the table $P[L, m, s]$ $(1 \leq L \leq m \leq k - s + 1,\ 1 \leq s \leq k)$.

We start by computing the formula (7.41) for $1 \leq m \leq 4$. Next, for $m \geq 5$, we perform the following computations with $m = 5,\ 6, \ldots,\ k$.

1. For $L = m$, using the table Q in Sect. 7.3.2, let

$$P[m, m, s] = Q\,[s, s+1, \ldots, s+m] \quad (1 \leq m \leq k - s + 1,\ 1 \leq s \leq k).$$

2. For $2 \leq L \leq m - 1$, compute

$$P[L, m, s] = P(L, m, (s, s+1, \ldots, s+m-1)]$$

by using the recurrence relation (7.43) with

$$\boldsymbol{w} = (w_s, w_{s+1}, \ldots, w_{s+m-1}).$$

Let $\boldsymbol{w}_i$ be a partial vector of the integer sequence, $s = s_0 < s_1 < \cdots < s_L = s + m$ of length $L + 1$ as follows:

$$\boldsymbol{w}_i = (w_{s_i}, w_{s_i+1}, \ldots, w_{s_{i+1}-1}) \quad (0 \le i \le L-1).$$

Then, since in view of the definition (7.39)

$$P(1, m_i, \boldsymbol{w}_i) = P[1, s_{i+1} - s_i, s_i],$$

we have

$$P[L, m, s] = \sum_{s=s_0<s_1<\cdots<s_L=s+m} Q\,[s_0, s_2, \ldots, s_L] \prod_{i=0}^{L-1} P\left[1, s_{i+1} - s_i, s_i\right],$$

where the summation is performed on the integer sequence, $s = s_0 < s_1 < \cdots < s_L = s + m$ of length $L + 1$. Note that the required values of $Q\,[s_0, s_1, \ldots, s_L]$ have been obtained in Sect. 7.3.2. Similarly, the values of $P\left[1, s_{i+1} - s_i, s_i\right]$ have already been obtained as well, since $s_{i+1} - s_i < m$.

3. For $L = 1$, compute

$$P[1, m, s] = 1 - \sum_{L=2}^{m} P[L, m, s] \quad (1 \le s \le k - m + 1).$$

Finally, we completed the table P.

7.3.4 Computation of Integrals

We explain the computation of the integrals in (7.48) and (7.49). The following theorem shows that the functions $f_{s_{m+1}}$ belong to the function family **G**. Therefore, the indefinite and definite integrals of the functions are efficiently computed by finite Sinc integral and the finite trapezoidal rule.

Theorem 7.13 *Let m and k be integers such that* $1 \le m \le k$. *For an integer sequence* $1 \le s_0 < s_1 < \cdots < s_m \le k+1$, *let* $\boldsymbol{s}_m = (s_0, s_1, \ldots, s_m)$. *Then, it follows that*

$$f_{\boldsymbol{s}_m}(x) \in \mathbf{G}\left(\sqrt{\frac{u_{m-1}}{2\pi}}, \frac{u_{m-1}}{2}, \frac{u}{2}\right), \tag{7.50}$$

$$u_{m-1} = \sum_{j=s_{m-1}}^{s_m-1} w_j, \; u = \sum_{j=s_0}^{s_m-1} w_j,$$

and for the indefinite integrals $F_{\boldsymbol{s}_m}(x)$,

$$0 < F_{\boldsymbol{s}_m}(x) \le \Phi\left(u_{m-1}^{-1/2}, x\right) < 1. \tag{7.51}$$

Recall that the level probability is invariant under the multiplication of the weight vector $\boldsymbol{w} = (w_1, w_2, \ldots, w_k)$ by a positive constant. So, in the sequel, for simplicity, we assume the computation with the normalized weight vector,

$$\sum_{i=1}^{k} w_i = 1. \tag{7.52}$$

Now, we set an equidistant sample points

$$\boldsymbol{x} = (x_j), \quad x_j = ih \quad (-n \leq j \leq n),$$

with the sample spacing $h > 0$ and the number $N = 2n + 1$ of sample points. The finite Sinc approximation and the finite Sinc integral of $f_{s_m}(x)$ on $\boldsymbol{x} = (x_j)$ are given by

$$c_{\boldsymbol{x}}^n[f_{s_m}](x) = \sum_{j=-n}^{n} f_{s_m}(x_j) s_h(x_j, x) \cong f_{s_m}(x),$$

$$C_{\boldsymbol{x}}^n[f_{s_m}](x) = \sum_{j=-n}^{n} f_{s_m}(x_j) S_h(x_j, x) \cong F_{s_m}(x),$$

respectively.

Here, we write the above two values in vector notation. Let the vectors of $f_{s_m}(x)$, $F_{s_m}(x)$, and $\varphi\left(u_{m-1}^{-1/2}, x\right)$ sampled on $\boldsymbol{x} = (x_j)$ be written by

$$\boldsymbol{f}_{s_m} = \left(f_{s_m}(x_j)\right)_{-n \leq j \leq n}, \quad \boldsymbol{F}_{s_m} = \left(F_{s_m}(x_j)\right)_{-n \leq j \leq n},$$

$$\boldsymbol{\varphi}_{u_{m-1}} = \left(\varphi\left(u_{m-1}^{-1/2}, x_j\right)\right)_{-n \leq j \leq n},$$

respectively. Define an integration matrix S by the square matrix of order $2n + 1$,

$$(S)_{i,j} = S_h(x_j, x_i) \quad (-n \leq i \leq n, \ -n \leq j \leq n). \tag{7.53}$$

Then, we have

$$F_{s_m}(x_i) \cong C_{\boldsymbol{x}}^n[f_{s_m}](x_i) = \sum_{j=-n}^{n} f_{s_m}(x_k) S_h(x_j, x_i)$$

$$= \sum_{j=-n}^{n} (S)_{i,j} f_{s_m}(x_j) \quad (-n \leq i \leq n),$$

or in vector notation,

$$\boldsymbol{F}_{s_m} \cong S \boldsymbol{f}_{s_m}.$$

From this relation, we compute (7.48) approximately by

$$\boldsymbol{f}_{s_m} = \boldsymbol{\varphi}_{u_{m-1}} \boldsymbol{F}_{s_{m-1}} = \boldsymbol{\varphi}_{u_{m-1}} \left(S \boldsymbol{f}_{s_{m-1}} \right), \tag{7.54}$$

where $\boldsymbol{f}\boldsymbol{g}$ denotes the element-wise product of the vectors $\boldsymbol{f}$ and $\boldsymbol{g}$. We compute the definite integral (7.49) by the finite trapezoidal rule

$$Q[\boldsymbol{s}_m] \cong T_x^n[f_{s_m}] = h \sum_{i=-n}^{n} f_{s_m}(x_i). \tag{7.55}$$

The main part of the computations is in (7.54) involving the matrix–vector product. Remark that $\boldsymbol{f}_{s_m}$ is a vector whose elements are function values at sample points of a quickly decreasing function. So, since the function values near both ends of the set of the sample points are very small in magnitude and have no contribution to the computation, we can reduce the amount of computations in (7.54) by deleting them and reducing the dimension of $\boldsymbol{f}_{s_m}$.

Finally, we describe how to determine the sample spacing h and the number of sample points $N = 2n + 1$. Let

$$w_{\min} = \min_{1 \le i \le k} w_i = \min_{1 \le i \le j \le k} \sum_{\ell=i}^{j} w_\ell.$$

Then, for an error criterion $\varepsilon > 0$, we recommend that the standard values of h and R (the half width of the approximate support $[-R, R]$) are

$$h = \sqrt{\frac{-\pi^2}{2 \log \varepsilon}}, \quad R = \sqrt{\frac{-2}{w_{\min}} \log \varepsilon}, \tag{7.56}$$

and

$$n = \lceil R/h \rceil, \tag{7.57}$$

where $\lceil \cdot \rceil$ is *the ceiling function* and $\lceil x \rceil$ is the least integer greater than or equal to the real number x.

In view of (7.50), we see that the minimum value of the Gauss parameter $\alpha = u_{m-1}/2$ of $f_{s_m}(x)$ is $\alpha_{min} = w_{min}/2$. Further, the maximum value of the Gauss parameter β is $\beta_{\max} = \sum_{\ell=1}^{k} w_\ell/2 = 1/2$.

Therefore, from Theorem 7.10 and the relations

$$\log \varepsilon = -\frac{\pi^2}{2h^2} = -\frac{w_{\min}}{2} R^2,$$

the discretization errors are

$$\left\| Ec_x[f_{s_m}] \right\|_\infty = O\left(e^{-\pi^2/(4\beta_{\max}h^2)}\right) = O\left(e^{-\pi^2/(2h^2)}\right) = O(\varepsilon),$$
$$\left\| EC_x[f_{s_m}] \right\|_\infty = O\left(e^{-\pi^2/(4\beta_{\max}h^2)}\right) = O\left(e^{-\pi^2/(2h^2)}\right) = O(\varepsilon),$$
$$\left| ET_x[f_{s_m}] \right| = O\left(e^{-\pi^2/(\beta_{\max}h^2)}\right) = O\left(e^{-2\pi^2/h^2}\right) = O\left(\varepsilon^4\right),$$

and the truncated errors are

$$\left\| \check{E}c_x^n[f_{s_m}] \right\|_\infty = O\left(e^{-\alpha_{\min}R^2}\right) = O\left(e^{-w_{\min}R^2/2}\right) = O(\varepsilon),$$
$$\left\| \check{E}C_x^n[f_{s_m}] \right\|_\infty = O\left(e^{-\alpha_{\min}R^2}\right) = O\left(e^{-w_{\min}R^2/2}\right) = O(\varepsilon),$$
$$\left| \check{E}T_x^n[f_{s_m}] \right| = O\left(e^{-\alpha_{\min}R^2}\right) = O\left(e^{-w_{\min}R^2/2}\right) = O(\varepsilon).$$

Then, we have

$$\left\| Ec_x^n[f_{s_m}] \right\|_\infty \le \left\| Ec_x[f_{s_m}] \right\|_\infty + \left\| \check{E}c_x^n[f_{s_m}] \right\|_\infty = O(\varepsilon),$$
$$\left\| EC_x^n[f_{s_m}] \right\|_\infty \le \left\| EC_x[f_{s_m}] \right\|_\infty + \left\| \check{E}C_x^n[f_{s_m}] \right\|_\infty = O(\varepsilon),$$
$$\left| ET_x^n[f_{s_m}] \right| \le \left| ET_x[f_{s_m}] \right| + \left| \check{E}T_x^n[f_{s_m}] \right| = O(\varepsilon).$$

We can estimate the error numerically by comparing the original calculation with more accurate one with a slightly smaller h and a slightly bigger R. We can also use this result to adjust h and R precisely.

7.3.5 *Numerical Experiments*

Using the algorithm mentioned above, we constructed computer programs in Mathematica and C language to perform numerical experiments on Macintosh iMac (3.2 GHz CPU and Intel Core i5).

The spacing of sample points used in Sect. 7.3.4 is $h = 0.8$.

[Experiment 1] We compute the table $P[L, m, s]$ of (7.39) for the weight vector

$$\boldsymbol{w} = (1, 1, 1, 1, 1, 1, 1, 1, 1, 1) \in \mathbb{R}^{10} \quad (k = 10),$$

with computation time, about 0.5 s in Mathematica and 0.003 s in C.

The computation in C is performed faster than in Mathematica by factor about 150, since Mathematica is an interpreter language and programs in C are compiled.

The maximum absolute error of $P[L, m, s]$ is 2.8×10^{-16} and the maximum relative error 1.6×10^{-14}. Since all weights are equal in this experiment, we compute the exact values (rational number) of $P[L, m, s]$ using the simple recurrence relation (7.40). These exact values enable us to estimate the errors of the computed results.

The number of elements in $P[L, m, s]$ is 220. We show among them the values of $P = P[L, 10, 1],\ 1 \leq L \leq 10$ as follows:

L	1	2	3	4	5	6	7	8	9	10
P	$\frac{1.000}{10}$	$\frac{2.829}{10}$	$\frac{3.232}{10}$	$\frac{1.994}{10}$	$\frac{7.422}{10^2}$	$\frac{1.744}{10^2}$	$\frac{2.604}{10^3}$	$\frac{2.397}{10^4}$	$\frac{1.240}{10^5}$	$\frac{2.756}{10^7}$

[Experiment 2] We compute the table $P[L, m, s]$ of (7.39) for the weight vector

$$\boldsymbol{w} = (1, 2, 3, 4, 5, 6, 7, 8, 9, 10) \in \mathbb{R}^{10} \quad (k = 10),$$

with computation time, about 3 s in Mathematica and 0.02 s in C.

Comparison of Experiment 1 and 2 reveals that the computation time required in this experiment is larger by factor 8. The required computation time is larger when the normalized minimum weight

$$w_{\min} = \frac{\min_{1 \leq i \leq k} w_i}{\sum_{i=1}^{k} w_i}$$

is smaller, since, from (7.57), the size of the matrix $S \in \mathbb{R}^{(2n+1)\times(2n+1)}$ of (7.53) is proportional to $1/\sqrt{w_{\min}}$.

The maximum absolute error of $P[L, m, s]$ computed is 3.4×10^{-16}. We estimate the error by the difference of the computed values and those with $h = 0.7$.

The number of elements in $P[L, m, s]$ is 220. We show among them the values of $P = P[L, 10, 1],\ 1 \leq L \leq 10$ as follows:

L	1	2	3	4	5	6	7	8	9	10
P	$\frac{7.717}{10^2}$	$\frac{2.442}{10^1}$	$\frac{3.189}{10^1}$	$\frac{2.280}{10^1}$	$\frac{9.906}{10^2}$	$\frac{2.731}{10^2}$	$\frac{4.802}{10^3}$	$\frac{5.215}{10^4}$	$\frac{3.186}{10^5}$	$\frac{8.366}{10^7}$

References

Abramowitz M, Stegun IA (1972) Handbook of mathematical functions. Dover, New York

Barlow RE, Bartholomew DJ, Bremner JM, Brunk HD (1972) Statistical inference under order restrictions. Wiley, London

Hayter AJ, Liu W (1996) On the exact calculation of the one-sided studentized range test. Comput Stat Data Anal 22:17–25

Lund J, Bowers KL (1992) Sinc methods for quadrature and differential equations. SIAM, Philadelphia

Ralston A, Rabinowitz P (1978) First course in numerical analysis, 2nd edn. Dover, New York

Robertson T, Wright FT, Dykstra RL (1988) Order restricted statistical inference. Wiley, New York

Shiraishi T (2011) Multiple comparison procedures under continuous distributions. Kyoritsu-Shuppan Co., Ltd. (in Japanese)

Shiraishi T, Sugiura H (2018) Theory of multiple comparison procedures and its computation. Kyoritsu-Shuppan Co., Ltd. (in Japanese)

Stenger F (1993) Numerical methods based on sinc and analytic function. Springer, Berlin

Takahashi H, Mori M (1974) Double exponential formulas for numerical integration. Publ Res Inst Math Sci 9:721–741

Chapter 8
Related Topics

Abstract Chapters 1–7 consider multiple comparisons for all-pairwise differences among mean responses in k samples under the assumptions of continuous distribution. This chapter discusses related topics including order restrictions, two-way layouts, and binomial responses.

8.1 Multiple Comparisons Under Simple Order Restrictions

We consider homoscedastic k sample models under normality of Sect. 3.1. Throughout this section, we use notations presented in Chap. 3. We add the assumption of simple order restrictions of (3.1).

For a specified i such that $1 \leq i \leq k-1$, if we are interested in a test of

$$\text{the null hypothesis } H_{(i,i+1)} : \mu_i = \mu_{i+1} \text{ vs. the alternative } H^{OA}_{(i,i+1)} : \mu_i < \mu_{i+1},$$

we can use a one-sided two-sample t-test based on
$T_i := \left(\bar{X}_{i+1\cdot} - \bar{X}_{i\cdot}\right) / \sqrt{V_E\left(1/n_{i+1} + 1/n_i\right)}$. We define $TD_2(t)$ as
$TD_2(t) := P_0\left(\max_{1 \leq i \leq k-1} T_i \leq t\right)$. For a given α such that $0 < \alpha < 1$, we put

$$td_2(k, n_1, \ldots, n_k; \alpha) := \text{a solution of } t \text{ satisfying the equation } TD_2(t) = 1 - \alpha.$$

By using $td_2(k, n_1, \ldots, n_k; \alpha)$, Lee and Spurrier (1995a) proposed single-step simultaneous tests for the successive comparisons of $\left\{\text{the null hypothesis } H_{(i,i+1)} \text{ vs. the alternative } H^{OA}_{(i,i+1)} \,\middle|\, 1 \leq i \leq k-1\right\}$. In addition, Lee and Spurrier (1995b) presented distribution-free single-step simultaneous tests based on rank statistics. Shiraishi (2014a) proposed two closed testing procedures which are superior to these single-step simultaneous tests of Lee and Spurrier (1995a, b).

Next, we consider multiple comparison tests for $\left\{\text{the null hypothesis } H_i : \mu_i = \mu_1 \text{ vs. the alternative } H_i^{OA} : \mu_i > \mu_1 \,\middle|\, 2 \leq i \leq k\right\}$. Williams (1971, 1972) proposed multistep test procedures under the following condition (C4) of sample sizes:

T.-a. Shiraishi et al., *Pairwise Multiple Comparisons*,
JSS Research Series in Statistics,
https://doi.org/10.1007/978-981-15-0066-4_8

$$(C4) \qquad n_2 = \cdots = n_k. \tag{8.1}$$

Williams (1971, 1972) used the statistics $T_\ell^o := \left(\tilde{\mu}_\ell^o - \bar{X}_{1\cdot}\right) / \sqrt{\left(\frac{1}{n_2} + \frac{1}{n_1}\right) V_E}$ $(\ell = 2, \ldots, k)$, where $\tilde{\mu}_\ell^o = \max_{2 \le s \le \ell} \left(\sum_{i=s}^{\ell} \bar{X}_{i\cdot}\right) / (\ell - s + 1)$. We define $TD_3(t|\ell, m, n_2/n_1)$ by $TD_3(t|\ell, m, n_2/n_1) := P_0\left(T_\ell^o \le t\right)$. For given α such that $0 < \alpha < 1$, we put

$$td_3(\ell, m, n_2/n_1; \alpha) := \text{a solution of } t \text{ satisfying the equation} \\ TD_3(t|\ell, m, n_2/n_1) = 1 - \alpha.$$

Williams (1971, 1972) used $td_3(\ell, m, n_2/n_1; \alpha)$ $(\ell = 2, \ldots, k)$ for the multistep test procedures.

We put $I_\ell^1 := \{1, 2, \ldots, \ell\}$ $(\ell = 2, \ldots, k)$. For $\ell = 2, \ldots, k$, we define $\tilde{\mu}_1^*(I_\ell^1), \ldots, \tilde{\mu}_\ell^*(I_\ell^1)$ by $u_1, \ldots, u_\ell$ which minimize $\sum_{i \in I_j} \lambda_{ni} \left(u_i - \bar{X}_{i\cdot}\right)^2$ under simple order restrictions $u_1 \le \cdots \le u_\ell$. That is,

$$\sum_{i=1}^{\ell} \lambda_{ni} \left(\tilde{\mu}_i^*(I_\ell^1) - \bar{X}_{i\cdot}\right)^2 = \min_{u_1 \le \cdots \le u_\ell} \sum_{i=1}^{\ell} \lambda_{ni} \left(u_i - \bar{X}_{i\cdot}\right)^2,$$

where λ_{ni} is defined by (3.2). Furthermore, we put $\bar{B}_3^2(I_\ell^1) := \sum_{i=1}^{\ell} n_i \left(\tilde{\mu}_i^*(I_\ell^1) - \bar{X}_{\cdot\cdot}(I_\ell^1)\right)^2 / V_E$, where $\bar{X}_{\cdot\cdot}(I_\ell^1) := \sum_{i=1}^{\ell} \sum_{j=1}^{n_i} X_{ij} / \sum_{i=1}^{\ell} n_i$. For given α such that $0 < \alpha < 1$, we put

$$\bar{b}_3^2(\ell, \boldsymbol{\lambda}_n(I_\ell^1), m; \alpha) := \text{a solution of } t \text{ satisfying the equation } P_0(\bar{B}_3^2(I_\ell^1) \ge t) = \alpha,$$

where $\boldsymbol{\lambda}_n(I_\ell^1) := (n_1/n, n_2/n, \ldots, n_\ell/n)$. Then Shiraishi and Sugiura (2018) proposed multiple comparison tests based on $\{\bar{B}_3^2(I_\ell^1),\ \bar{b}_3^2(\ell, \boldsymbol{\lambda}_n(I_\ell^1), m; \alpha) \mid \ell = 2, \ldots, k\}$ for $\left\{\text{the null hypothesis } H_i \text{ vs. the alternative } H_i^{OA} \,\middle|\, 2 \le i \le k\right\}$. The multiple comparison procedure proposed by Shiraishi and Sugiura (2018) need not impose the condition (C4). By the way similar to Chap. 7, we can compute $td_2(k, n_1, \ldots, n_k; \alpha)$, $td_3(\ell, m, n_2/n_1; \alpha)$ and $\bar{b}_3^2(\ell, \boldsymbol{\lambda}_n(I_\ell^1), m; \alpha)$. The theory of the numerical computation is stated in Shiraishi and Sugiura (2018).

8.2 Two-Way Layouts

We consider a randomized block design with n blocks and k treatments. The commonly used model for the randomized block design is

$$X_{ij} = \mu + \beta_i + \tau_j + e_{ij} \quad (1 \le i \le n,\ 1 \le j \le k), \tag{8.2}$$

where $\sum_{i=1}^{n} \beta_i = \sum_{j=1}^{k} \tau_j = 0$ and the e_{ij}'s are independent and normally distributed with mean 0 and unknown variance σ^2. μ, β_i, and τ_j are referred to as the overall mean, the ith block effect and the jth treatment effect, respectively. The least squares estimators of μ, β_i, and τ_j, respectively, are $\hat{\mu} = \bar{X}_{\cdot\cdot}$, $\hat{\beta}_i = \bar{X}_{i\cdot} - \bar{X}_{\cdot\cdot}$, and $\hat{\tau}_j = \bar{X}_{\cdot j} - \bar{X}_{\cdot\cdot}$, where $\bar{X}_{\cdot\cdot} = \{1/(nk)\} \sum_{i=1}^{n} \sum_{j=1}^{k} X_{ij}$, $\bar{X}_{i\cdot} = (1/k) \sum_{j=1}^{k} X_{ij}$, and $\bar{X}_{\cdot j} = (1/n) \sum_{i=1}^{n} X_{ij}$. An unbiased estimator of σ^2 is given by $\hat{\sigma}^2 = \sum_{i=1}^{n} \sum_{j=1}^{k} (X_{ij} - \bar{X}_{i\cdot} - \bar{X}_{\cdot j} + \bar{X}_{\cdot\cdot})^2 / m_1$, where $m_1 := (n-1)(k-1)$. We put $T_{j'j}(\boldsymbol{\tau}) := (\hat{\tau}_{j'} - \hat{\tau}_j - \tau_{j'} + \tau_j)/\sqrt{2\hat{\sigma}^2/n}$. We then find

$$P\left(\max_{1 \le j < j' \le k} |T_{j'j}(\boldsymbol{\tau})| \le t\right) = TA(t|k, m_1), \tag{8.3}$$

where $TA(t|k, m_1)$ is given by replacing ℓ and m with k and m_1, respectively, in (1.20). From the relation of (8.3), Tamhane (2009) pointed out that we can derive Tukey-type simultaneous confidence intervals for all-pairwise comparisons of $\{\tau_{j'} - \tau_j \mid (j, j') \in \mathscr{U}\}$, where $\mathscr{U}$ is defined by (1.5). Similarly, we can give Tukey-type simultaneous tests for all-pairwise comparisons of {the null hypothesis $H^t_{(j,j')}$: $\tau_j = \tau_{j'}$ vs. the alternative hypothesis $H^{tA}_{(j,j')}$: $\tau_j \neq \tau_{j'} \mid (j, j') \in \mathscr{U}$}. Shiraishi and Matsuda (2018) proposed closed testing procedures based on the maximum values of some $T_{j'j}$'s. In this paper, it is shown that the proposed procedures are more powerful than the single-step Tukey-type tests and the REGW (Ryan–Einot–Gabriel–Welsch)-type tests. Furthermore, Shiraishi and Matsuda (2018) discussed multiple comparison procedures in the model (8.2) under simple order restrictions of $\tau_1 \le \tau_2 \le \cdots \le \tau_k$.

Next, in model (8.2), we change the assumption that the e_{ij}'s are independent and identically distributed with the absolutely continuous distribution function $F(x)$ such that $\int_{-\infty}^{\infty} x dF(x) = 0$. Shiraishi and Matsuda (2019b) proposed closed testing procedures based on signed rank statistics and Friedman test statistics for all-pairwise comparisons of treatment effects. Although anyone has been failed to discuss a distribution-free method except Bonferroni procedures as a multiple comparison test, the proposed procedures are exactly distribution-free. Next, we consider the randomized block design under simple ordered restrictions of treatment effects. We propose distribution-free closed testing procedures based on one-sided signed rank statistics and rank statistics of Chacko (1963) for all-pairwise comparisons. Simulation studies are performed under the null hypothesis and some alternative hypotheses. In these studies, the proposed procedures show a good performance. We also illustrate an application to death rates by using proposed procedures.

Finally, we consider a two-way ANOVA with interaction. For the two-way model, the kth observation X_{ijk} in the ith level of the first factor and jth level of the second factor is expressed as

$$X_{ijk} = \mu + a_i + b_j + (ab)_{ij} + e_{ijk} \quad (i = 1, \ldots, I;\ j = 1, \ldots, J;\ k = 1, \ldots, n), \tag{8.4}$$

where $\sum_{i=1}^{I} a_i = \sum_{j=1}^{J} b_j = 0, \sum_{i=1}^{I}(ab)_{ij} = 0\,(j = 1, \ldots, J)$, and $\sum_{j=1}^{J}(ab)_{ij} = 0\ (i = 1, \ldots, I)$. In (8.4), μ is the overall mean, a_i is the effect of ith level of of the first factor and the jth level of the second factor, $(ab)_{ij}$ is the interaction between the ith level of of the first factor and the jth level of the second factor, and e_{ijk} is the error term. It is assumed that e_{ijk}'s are independent and normally distributed with mean 0 and unknown variance σ^2. The estimators of a_i, b_j, $(ab)_{ij}$, and σ^2 are, respectively, given by $\widetilde{a}_i = \bar{X}_{i..} - \bar{X}_{...}$, $\widetilde{b}_j = \bar{X}_{.j.} - \bar{X}_{...}$, $\widetilde{(ab)}_{ij} = \bar{X}_{ij.} - \bar{X}_{i..} - \bar{X}_{.j.} + \bar{X}_{...}$, and $\widetilde{\sigma}^2 = \sum_{i=1}^{I}\sum_{j=1}^{J}\sum_{k=1}^{n}(X_{ijk} - \bar{X}_{ij.})^2/m_2$, where $\bar{X}_{ij.} = \sum_{k=1}^{n} X_{ijk}/n$, $\bar{X}_{i..} = \sum_{j=1}^{J}\sum_{k=1}^{n} X_{ijk}/(Jn)$, $\bar{X}_{.j.} = \sum_{i=1}^{I}\sum_{k=1}^{n} X_{ijk}/(In)$, and $m_2 = IJ(n-1)$. Let us put $T_{i'i}(\boldsymbol{a}) := \sqrt{Jn}(\widetilde{a}_{i'} - \widetilde{a}_i - a_{i'} + a_i)/\sqrt{2\widetilde{\sigma}^2}$. Then we have the following equation similar to (8.3):

$$P\left(\max_{1 \le i < i' \le I} |T_{i'i}(\boldsymbol{a})| \le t\right) = TA(t|I, m_2).$$

Thus, we can derive Tukey-type simultaneous confidence intervals for all-pairwise comparisons of $\{a_{i'} - a_i \mid 1 \le i < i' \le I\}$. Similarly, we can give Tukey-type simultaneous tests for all-pairwise comparisons of {the null hypothesis $H^a_{(i,i')} : a_i = a_{i'}$ vs. the alternative hypothesis $H^{aA}_{(i,i')} : a_i \ne a_{i'} \mid 1 \le i < i' \le I\}$. Shiraishi and Matsuda (2019a) proposed closed testing procedures. Furthermore, Shiraishi and Matsuda (2019a) discuss multiple comparison procedures in model (8.4) under the simple order restrictions of $a_1 \le a_2 \le \cdots \le a_I$.

8.3 Bernoulli Responses

We consider k-sample models with Bernoulli responses. Specifically, suppose that $(X_{i1}, \ldots, X_{in_i})$ is a random sample of size n_i from the ith Bernoulli population with unknown success probability p_i $(i = 1, \ldots, k)$. Furthermore, X_{ij}'s are assumed to be independent. Then $X_i = \sum_{j=1}^{n_i} X_{ij}$ has the binomial distribution with parameters n_i and p_i, and the estimators of p_i's are given by $\hat{p}_i = X_i/n_i$ or $(X_i + 0.5)/(n_i + 1)$ $(i = 1, \ldots, k)$. Using a central limit theorem and Slutsky's theorem, under condition (C1) of (2.5), we get

$$\sqrt{n}\left\{2\arcsin\left(\sqrt{\hat{p}_i}\right) - 2\arcsin\left(\sqrt{p_i}\right)\right\} \xrightarrow{\mathscr{L}} Y_i \sim N\left(0, \frac{1}{\lambda_i}\right), \tag{8.5}$$

where $n := \sum_{i=1}^{k} n_i$ and $Y \sim N(0, \sigma^2)$ denotes that Y is distributed according to $N(0, \sigma^2)$. For $i < i'$, let us put

$$T_{i'i}^{B} := \frac{2\left\{\arcsin\left(\sqrt{\hat{p}_{i'}}\right) - \arcsin\left(\sqrt{\hat{p}_{i}}\right)\right\}}{\sqrt{\frac{1}{n_i} + \frac{1}{n_{i'}}}}.$$

Then from (8.5), under (C1) and the condition of $p_1 = \ldots = p_k$, we get

$$A(t) \leq \lim_{n\to\infty} P_0\left(\max_{1\leq i<i'\leq k} |T_{i'i}^{B}| \leq t\right) \leq A^*(t),$$

where $A(t)$ and $A^*(t)$ are defined in (4.5). Thus, the Tukey–Kramer-type simultaneous test of level α for all-pairwise comparisons of
$\left\{\text{the null hypothesis } H_{(i,i')} : p_i = p_{i'} \text{ vs. the alternative } H_{(i,i')}^{A} : p_i \neq p_{i'} \,\middle|\, (i,i') \in \mathcal{U}\right\}$
consists in rejecting $H_{(i,i')}$ for $(i,i') \in \mathcal{U}$ such that $|T_{i'i}^{B}| > a(k;\alpha)$, where $a(k;\alpha)$ is defined by (2.10). Shiraishi (2011) proposed the closed testing procedure based on $|T_{i'i}^{B}|$'s which is superior to the Tukey–Kramer-type simultaneous test. Furthermore, Shiraishi (2014b) and Shiraishi and Matsuda (2016) discussed multiple comparison procedures in the k-sample models with Bernoulli responses under the simple order restrictions of $p_1 \leq p_2 \leq \cdots \leq p_k$.

References

Chacko VJ (1963) Testing homogeneity against ordered alternatives. Ann Math Statist 34:945–956

Lee RE, Spurrier JD (1995a) Successive comparisons between ordered treatments. J Statist Plann Infer 43:323–330

Lee RE, Spurrier JD (1995b) Distribution-free multiple comparisons between successive treatments. J Nonparametric Statist 5:261–273

Shiraishi T (2011) Multiple tests based on arcsin transformation in multi-sample models with Bernoulli responses. Jpn Soc Appl Statist 40:1–17 (in Japanese)

Shiraishi T (2014a) Closed testing procedures in multi-sample models under a simple order restriction. J Jpn Stat Society Jpn Issue 43:215–245 (in Japanese)

Shiraishi T (2014b) Multiple comparison procedures for a simple ordered restriction in multi-sample models with Bernoulli responses. Jpn Soc Appl Statist 43:1–22 (in Japanese)

Shiraishi T, Matsuda S (2016) Closed testing procedures based on $\bar{\chi}^2$-statistics in multi-sample models with Bernoulli responses under simple ordered restrictions. Jpn J Biom 37:67–87

Shiraishi T, Matsuda S (2018) Closed testing procedures for all pairwise comparisons in a randomized block design. Commun Stat-Theory Methods 47:3571–3587

Shiraishi T, Matsuda S (2019a) Closed testing procedures for all pairwise comparisons of main effects in the two-way layout with replications. J Jpn Stat Society Jpn Issue 48. (Accepted for publication)

Shiraishi T, Matsuda S (2019b) Nonparametric closed testing procedures for all pairwise comparisons in a randomized block design. Biom Soc Jpn 40:1–14

Shiraishi T, Sugiura H (2018) Theory of multiple comparison procedures and its computation. Kyoritsu-Shuppan Co., Ltd (in Japanese)

Tamhane AC (2009) Statistical analysis of designed experiments. Wiley, New York

Williams DA (1971) A test for differences between treatment means when several dose levels are compared with a zero dose control. Biometrics 27:103–117
Williams DA (1972) The comparison of several dose levels are compared with a zero dose control. Biometrics 28:519–531

Index

T.-a. Shiraishi et al., *Pairwise Multiple Comparisons*,
JSS Research Series in Statistics,
https://doi.org/10.1007/978-981-15-0066-4

Printed in the United States
By Bookmasters